MERE
SCIENCE
and
CHRISTIAN
FAITH

BRIDGING THE
DIVIDE WITH
EMERGING ADULTS

GREG COOTSONA

IVP Books
An imprint of InterVarsity Press
Downers Grove, Illinois

InterVarsity Press
P.O. Box 1400, Downers Grove, IL 60515-1426
ivpress.com
email@ivpress.com

InterVarsity Press® is the book-publishing division of InterVarsity Christian Fellowship/USA®, a movement of students and faculty active on campus at hundreds of universities, colleges, and schools of nursing in the United States of America, and a member movement of the International Fellowship of Evangelical Students. For information about local and regional activities, visit intervarsity.org.

While any stories in this book are true, some names and identifying information may have been changed to protect the privacy of individuals.

Cover design: David Fassett
Interior design: Jeanna Wiggins
Images: color spheres: © Iconica / Getty Images
    mountains: © cdbrphotography /iStockphoto / Getty Images
    night sky: © michal-rojek / iStockphoto / Getty Images

ISBN 978-0-8308-3814-1 (print)
ISBN 978-0-8308-8741-5 (digital)

Printed in the United States of America ∞

Library of Congress Cataloging-in-Publication Data
A catalog record for this book is available from the Library of Congress.

P	25	24	23	22	21	20	19	18	17	16	15	14	13	12	11	10	9	8	7	6	5	4	3	2	1
Y	37	36	35	34	33	32	31	30	29	28	27	26	25	24	23	22	21	20	19	18					

THIS BOOK IS DEDICATED TO

*those friends and colleagues who have discussed
these topics over the past several decades.*

*I appreciate your wit, insight,
and indulgence.*

CONTENTS

1

CREATION, BEAUTY, and SCIENCE

*The world doesn't want to mention both
religion and science in the same sentence.
But it shouldn't be that way.*

ELIANA, AGE 18

One summer not too long ago, when my daughter Melanie was about to head to college, I found myself on a backpacking trip with her, a number of other high schoolers from our church, and their parents. We were making our way through the lower reaches of the Sierra Nevadas, where stunning, speckled-greyish granite filled our view. Our group of about sixteen had just left the beauties of California Highway 20 as it wound its way past the former mining town of Nevada City and climbed for a couple dozen miles through glorious pine forests. We arrived at a dirt road, full of divots and ruts, which we drove for several miles until we arrived at a parking lot near the trailhead. It was a warm summer

Sierra morning with the hint of a few clouds. Backpacks filled with tents, sleeping bags, freeze-dried food, a few clothes, toiletries, and the like, we began our hike toward the Five Lakes Basin.

I thought the primary point of the trip was to get to the camping spot on Grouse Ridge. But on the way, Ella, the seven-year-old daughter of the trip leader and youth pastor, taught me something surprising. As we worked our way up the trail, she found beautiful rocks and brought them to her father, who had a degree in geology.

"Aren't these amazing? Isn't this one really pretty, Dad?"

He affirmed her discoveries and dutifully collected each of her finds (which gradually filled and weighed down his backpack). Once a rock was deposited, Ella would run off and search.

"Here's another one, Dad! Look at the colors!" It was an implicit and exuberant celebration of the goodness of creation. "This one's even prettier!"

Ella's exclamations about geological beauty probably don't sound as significant as a fresh proposal for unifying quantum and relativity theories, but later I remarked to myself, "What Ella did—that's the beginning of science."

I later recalled a science and theology conference, particularly a discussion with a biologist who simply stated, "I find biological science fascinating and have ever since I was young. In fact, every scientist I know began with a profound experience with nature as a child."

It was my Outdoor Education class in fifth grade, when my classmates and I discovered banana slugs among the redwoods along moist dirt paths in the Santa Cruz Mountains. It was a chemistry set my friend received as a present for her eleventh birthday that helped her begin to find the joys of how chemicals interact. It was "just being fascinated as a kid by the bugs in the cracks of the sidewalk—I simply had to study them for hours," a PhD student in biochemistry said when describing the beginning of his life as a scientist.

As Ella's surprise and discovery that summer showed, the love of the surprisingly intricate beauty of nature is the beginning of science. This experience is intuitive in our early years of life—God created the world; let's figure out how to understand it.

Kids know this, but they're not the only ones who do. The great astronomer Nicolaus Copernicus observed, "The universe has been wrought for us by a supremely good and orderly Creator." That order led Copernicus to study the organization of the solar system and to rethink the configuration of the planets with our sun. (Enter heliocentrism.) Copernicus began with a conviction that God's good creation invites us to investigate. In our day, Francis Collins, the head of the National Institutes of Health—and also a follower of Jesus—once commented, "I find that studying the natural world is an opportunity to observe the majesty, the elegance, the intricacy of God's creation." And though Collins is quite clear that nature can't prove God's existence, studying God's works leads him—and us—to recognize fingerprints of God the Worker.

Collins, Copernicus, and Ella all agree. And they've got the Bible on their side: we are created to relate to the creation around us. The thrill scientists experience is a sense of excitement at discovering the natural world. In this sense the psalmist was acting as a natural scientist when he exclaimed,

What a wildly wonderful world, GOD!
> You made it all, with Wisdom at your side,
> made earth overflow with your wonderful creations.
>> (Psalm 104:24, *The Message*)

While the study of nature is the beginning of science, it's also a call to all believers. Although "the heavens declare the glory of God; the skies proclaim the work of his hands" (Psalm 19:1), many of us have become dulled to nature's divine speech, and scientists help us tune our ears to the mystery of a starlit night, the sophisticated order of our bodies, and the glorious structures of physical systems. In a graduate seminar on theology and science, I listened to a Berkeley biochemist describe the formations of polymers to the theologians. (Full disclosure: until that moment I had never carefully observed polymers.) He showed us a magnified picture, and in the midst of his careful explanation, he couldn't help exclaiming, "Look how beautiful these are!" After forty years of university teaching, his wonder and excitement were no less profound than Ella's.

When we grasp beauty in nature, we are drawn to the source of beauty. And the nature of beauty is that it draws us in. I'm reminded that in Eastern Orthodoxy, theology begins with *philokalia*, or "the love of beauty."[1] It also brings to mind the great Puritan pastors of the eighteenth century like Jonathan Edwards, who studied both nature and Scripture as sources for finding beauty. Edwards wrote,

> For as God is infinitely the greatest being, so he is allowed to be infinitely the most beautiful and excellent: and all the beauty to be found throughout the whole creation is but the reflection of the diffused beams of that Being who hath an infinite fullness of brightness and glory.[2]

Finally, I think of the lyrics from an alternative rock band (one of my favorites), Future of Forestry: "I will go where beauty leads me home." In this home we find God. Science thus leads us to grasp both the mystery and the majesty of creation. Jeff Hardin, zoologist at the University of Wisconsin, Madison, summarized it well at a recent BioLogos science and Christian faith conference: "Why be a scientist? Worship."

With this book I want to inspire more ministry leaders to point emerging adults toward studying nature as an act of worship. And it's not difficult. We love the natural world because God created it. And with the book of Genesis, we celebrate that this world is "very good" (Genesis 1:31). We see in and through this world the good God we know in Jesus Christ. All that is good news. And we who don't practice science as a profession but who seek to know God more deeply need to listen to scientists.

The Crooked Path of Nature

Through that listening we find praise and wonder and mystery. Scientists have also taught me honesty and a somewhat recalcitrant commitment to avoid easy answers by pondering intricacies we would have never guessed. (This may be why, in fact, some believers resist science—because scientists resist easy answers.) "Consider what God has done," Ecclesiastes 7:13 says. "Who can straighten what he has made crooked?" Sometimes the works of God in the ways of nature are not as straightforward as we would like, even though science has figured out numerous things ancient thinkers and New Testament writers didn't know. Nevertheless, through all this beauty, awesome display, and puzzling natural reality, we still somehow discover the "eternal power and divine nature" of our Creator (Romans 1:20). It strikes me that affirming the "eternal power and divine nature" of God offers both a wide place for scientific discovery and a respectful silence and patience for future answers. I believe that scientists ultimately lead us to admit our limits and declare the majesty of God, echoing what Paul exclaimed ten chapters later: "Oh, the depth of the riches of the wisdom and knowledge of God! How unsearchable his judgments, and his paths beyond tracing out!" (Romans 11:33).

We in the church need scientists to lead us to appreciate God's creation, which is our call as Christians. That's one major reason to "bring science to church"—to bring together our faith in Christ and the insights and discoveries of science.[3]

The Complications Begin

At some point, often after early childhood, this process of enjoying nature (and thus science) while believing in God (which most of us do naturally in our early childhood, as research by the cognitive scientist Justin Barrett has demonstrated[4]) becomes unwound by the voices around us.

Maybe we hear, like my daughter did, that bringing science and belief together is flat-out impossible. One day at Marigold Elementary School in Chico, California, parents came and described their careers to eager fourth-graders. That night at dinner Laura and I asked Lizzie, "How was career day?"

"Mom and Dad, it was weird. We had this scientist come to class— Brin's mother—and she basically told us that you couldn't be religious and be a scientist."

Naturally, I found this declaration a bit alarming and first channeled the empathic listening skills of my pastoral training. "Lizzie, I'm sorry to hear that—it must have been difficult."

"Well, it was uncomfortable, and it made me wonder about the Bible and science."

Now I couldn't restrain myself. "This has been my area of focus in work for the past two decades, and I can tell you that there are a number of great scientists who are Christians and many church leaders who accept the conclusions of science."

Crisis averted? Perhaps. But many people also hear in church that combining science and faith is impossible. Recently I had a conversation with an astronomer who told me a tragic story. James was brilliant in science and later a Rhodes Scholar, and he had grown up in a church that rejected the consensus of science. His pastors declared simply that modern mainstream science and human evolution were false (even though this paradigm has been tried and tested for over

150 years). As James began to study mainstream science, he realized how hard scientists had worked to derive their theories and how well it all fit together. This caused an astoundingly poignant psychological crisis, one so severe that in college James simply could not keep his faith without losing his mind. "The pastor taught me absolute gibberish about science," he said. "So how I can believe what he told me about the Word of God?"

Listening to ancient voices reminds me how old this problem is. Augustine (354–430), perhaps the most influential Christian thinker of antiquity, believed we should not speak rashly, even out of ignorance. Instead we need to engage the best science of our day or we mar our witness to the gospel. Here's why:

> If they [people outside the church] find a Christian mistaken in a field which they themselves know well and hear him maintaining his foolish opinions about our books, how are they going to believe those books in matters concerning the resurrection of the dead, the hope of eternal life, and the kingdom of heaven, when they think their pages are full of falsehoods and on facts which they themselves have learnt from experience and the light of reason?[5]

To be sure, the best science of Augustine's day was a universe of several thousand years. But we have better, more complete science, and if we talk about a six-thousand-year-old earth today, that sounds like nonsense to those outside the church—and most inside as well. (More on that in chapter four.) So let's not fool around with science that cannot be supported by scientists. This is a theme that we'll revisit throughout this book.

Christians have every reason, therefore, not to twist science into their own convenient configurations. In a speech to Anglican youth workers, C. S. Lewis noted that "science twisted in the interests of apologetics would be a sin and a folly."[6] This means the church needs scientists. They keep us honest, helping us avoid superstition and error. One emerging adult–ministry leader in Silicon Valley told me, "We are committed to teach things about science here in church that you would hear there," and he pointed to Stanford University. This particularly involved Big Bang cosmology, the critical and inextricable relationship between our brain and our mind, and the theory of evolution. As Lewis also wrote, "With Darwinianism as a theorem in Biology I do not think a Christian need have any quarrel."[7]

Whatever human knowledge discovers in nature, we are bound to listen, to learn, and to engage with it. Why? Because God has spoken and continues to speak through Scripture and through the natural world—through both words and works—albeit in different modes. Faith and science are not in a wrestling match where one will be the victor. In fact, Christians throughout the ages have celebrated that the same God who is visible in science is revealed in the pages of the Bible.

This points to an important axiom that we'll continue to explore: God speaks through "two books"—the book of nature and the book of Scripture—and these two books do not contradict one another. They have different, nonredundant messages. We learn through special revelation that God has acted and spoken to Israel and supremely in Jesus Christ. In special revelation we hear the definitive message of grace and salvation. In general revelation, God is still revealing himself but in more "general" ways. In creation we see God's "invisible qualities—his eternal power and divine nature" (Romans 1:20). God created nature of his free design, and thus we are compelled to study it. In addition, God has given us the ability as human beings to comprehend what he has created. As Francis Bacon, one of the pioneers of modern science, phrased it, "God has, in fact, written two books, not just one. Of course, we are all familiar with the first book he wrote, namely Scripture. But he has written a second book called creation."[8] We interpret these books in light of their contents.

Christians therefore have few reasons to be antiscience.

Emerging Adults, the Church, and Science

Many churches fail to treat the topic of science at all, even as their high schoolers are trying to put their faith together with what they learn about the natural world in the classroom. In his study of emerging adults, Barna president and researcher David Kinnaman discovered that 52 percent of youth group members will ultimately enter a science-related profession, but only 1 percent of youth groups talk about science even once a year.[9] Having spent about two decades in college ministry, I doubt the statistics in college groups are much different.

Many of us hear the story that science and Christianity aren't compatible, and it's one reason that emerging adults in the critical years between eighteen and thirty aren't affiliating with any Christian congregation. And that's why this book is *necessary*.

Consider the fact that the emerging adult generation is estimated at eighty to ninety million people. When asked the question, "Which religion do you affiliate with?" about a third of eighteen- to thirty-year-olds answer, "None."[10] This label has stuck—those not affiliated with any religious tradition are often called "nones." Kinnaman found that one of the top reasons emerging adults leave the church is they identify it as "antiscience."[11] Another primary reason these individuals abandon faith is that the church presents itself as having all the answers, which strikes me as related; many of the church's "answers" about science are

> "To be honest, I think that learning about science was the straw that broke the camel's back. I knew from church I couldn't believe in both science and God, so that was it. I didn't believe in God anymore."
>
> MIKE, QUOTED IN DAVID KINNAMAN, YOU LOST ME

what // says that -

incorrect. This means that a large proportion of the reasons given for rejecting the faith stem from the church's rejection of science.

If Kinnaman is right, unless Christian congregations work to bring science back into church, there may be millions fewer people in American pews in the coming years, and ultimately there may a visibly diminished church left to engage science. I'm not arguing that we should integrate faith with mainstream science just to gain converts— though I think that will happen. Rather, I'm convinced that the church must do the work of integration because if we don't, we throw away our legacy of Christians' contribution to natural science, like Copernicus's discovery at the dawn of the sixteenth century.

We have science as a birthright in the church and love science at its best because it discovers truth. And the Christian church is at its best when it seeks truth. I'll return to these themes, but first I need to describe how this narrative of faith and science played out in my college years.

Go to Berkeley. Become a Christian.

I was bottle-fed on the casual, happy secularism of the region now known as Silicon Valley. I grew up not needing God because I was satisfied by superb weather, comfortable surroundings, and a sufficient degree of personal achievement. If I had a theological creed, it was agnosticism or functional atheism. To be clear, I was a buoyant, secular Northern Californian, not some kind of dour, atheistic postmodernist. To me—and many around me—the existence of a deity didn't seem relevant or advantageous. So it was easy for me to wander the path set by a self-sufficient San Mateo County, where just a small percentage of residents could be caught in church on any given Sunday. Recently, the Barna Group found Oakland–San Francisco–San Jose to be the number-one "unchurched" and "dechurched" region in the country, followed closely by Chico-Redding, where I live now.[12] Though the title didn't exist at the time, I now count my early self as a "none." And I can't deny that some residual "noneness" still flows through my veins. It's certainly still the cultural air I breathe.

At age seventeen I started at the University of California, Berkeley, and shortly thereafter became a follower of Christ. I admit it: "Grow up in a secular home. Go to Berkeley. Become a Christian"—it's almost laughable. But that's what happened.

As a first-year student I was dazed by this spectacular university and undone by my newfound collegiate license. No parent or teacher could provide me with new certainties, and, quite frankly, the old ones didn't work so well. The voice of self-sufficiency, Ayn Rand's "virtue of self-ishness" (which I had learned at home), rang hollow, and so did whatever personal fulfillments I could cobble together.

Admittedly, this search for God wasn't purely intellectual—I've since learned that we don't engage arguments in abstraction; we engage with people we respect. But it wasn't anti-intellectual either. I found stunningly respectable, intelligent Christians. We had arguments, conversations, and more arguments in fraternities, at Berkeley cafes, and beside lockers. My best friend, Mike—the kind of brilliantly articulate antagonist who receives top scores on his advanced placement exams and later studies at Berkeley Law to become a prosecuting attorney—tried to argue me out of every nascent theological affirmation.

"How can you believe this Bible?"

"Do you think I'm going to hell?"

"Okay, explain to me how all this God stuff makes sense in light of science."

These friends handed me various books, many now forgotten, with the exception of the Bible and works by C. S. Lewis. *Mere Christianity*, which describes Lewis's intellectual disenchantment with atheism, got under my skin with its reasoned and reasonable approach to Christian faith. Lewis's approach, by the way, also taught me that Christian thought could engage any cultural influence—including science.

In the second quarter of my first year in this exquisitely secular college, without every answer clearly figured out, I committed my life to following Jesus. These days I often find myself balanced between a cradle secularism and a practiced faith. Even though it's been over three decades now, it's not hard for me to imagine the mindset of nones. They look for anything that seems to deny God's existence—and for many, science does a satisfactory job.

I also heard at Cal that there's no way to put faith and science together. I had invited one of my favorite professors, a visiting scholar from Germany named Friederike Haussauer, to have dinner with my parents, who were visiting from Menlo Park, about fifty miles south. As we sat together at Upstart and Crow Café on Berkeley's Bancroft Avenue, discussing various topics about Germany and the States (my mother's side of the family is German, so the motherland was a topic of common interest), Dr. Haussauer heard an incidental remark that I believed in God.

She cut right to the chase with what I'll call the Haussauer Problematic: "What possible sense does that make after modern science and the Enlightenment?" she asked. "How could you believe in God after Hume and Kant?"

Right in the middle of our quiches and hamburgers! A bit stunned, I had little to say. The conversation continued, and later we offered our goodbyes. Later, in her class on Enlightenment literature and thought, we read Voltaire, d'Alembert, and the other philosophes, who joined their French voices to my Deutsch professor's—true intellectuals have concluded that science presents decisive reasons for not believing in God.

I wish this were simply my experience. But over the past few years I've interviewed a number of students who've experienced similar antagonism from students and professors. Kelly, at twenty-one, experienced specific attacks. "I often find myself the only Christian in science class," she told me. "At one point a student blurted out, 'Christians are stupid.' My response: 'Do you think I'm stupid?' There was no response. On another occasion someone said, 'You cannot be Christian and scientist.'"

"It is almost always a topic of science versus religion, rarely a topic of science and religion."
ANDY, AGE 20

My Angle on Science and Faith

When I mention that I specialize in religion and science, one of the first questions I get is, "Are you a scientist?" And as you probably can tell, I'm not. Nonetheless, there's a reason I'm fascinated, and I'm willing to enter the conversation. I'm a Christian—as well as a former pastor, and now an academic specializing in this field of science and religion—who has found that scientific insights enhance, challenge, and strengthen my faith. Science emerges for most people as a way to make sense of things, as a worldview. So, to be a bit technical, I'm creating a theology of culture with science as a key component of that culture. By "culture," I mean what we humans make and add to the natural world—technically, the collective contributions of human intellectual and creative output.

To be clear, it's not hard to find people of deep faith and glittering scientific credentials. (This of course offers powerful counter-evidence to the story that all scientists hate religion.) One of my tasks over the past three decades has been to talk with as many of them as I can. For example, a few years ago I participated in a panel discussion at an exhilarating conference sponsored by DoSER—Dialogue on Science, Ethics, and Religion—a part of the American Association for the Advancement of Science (AAAS), the largest science organization in the world. The panel, titled "Perceptions: Science and Religious Communities," brought together evangelical Christians and scientists for conversation with the likes of Nobel Laureate physicist William D. Phillips, NASA astrophysicist Jennifer Wiseman (also head of DoSER), and Rice University sociologist Elaine Howard Ecklund, who surveyed over ten thousand Americans and found that that nearly 70 percent of self-identified evangelicals (often viewed as antiscience) do not see religion and science as being in conflict.[13] When interviewed by *National Geographic*, NIH's Francis Collins commented, "Science and faith can actually be mutually enriching and complementary once their proper domains are understood and respected."[14]

Huh?

I've discovered that for the Christian message to have any impact today, it must engage science. To appropriate the term Lewis made famous, *mere Christianity* needs to meet mainstream science. That's why I've focused much of my life's work on science and faith. Moreover, since I'm trained in theology and biblical studies, I can help sort out whether scientific insights and assertions are neutral, helpful, or antithetical to our faith. Why? Because it's not simply about scientists having difficulty with faith but also about people of faith not being sure about science. Is science, and particularly evolution, the "universal acid" that philosopher Daniel Dennett has warned us about? I'm convinced that's hyperbole.

That's why I hope a tribe of nonscientists will bring science to church. Let's not leave that task just to the science specialists. I often cite Pope John Paul II from his letter to astronomer George Coyne, former head of the Vatican Observatory: "Science can purify religion from error and superstition. Religion can purify science from idolatry and false absolutes."[15]

Terms and Definitions

"The second I say I'm a Christian, I get dismissed as unintelligent."

CHELSEA, AGE 20

I've learned the hard way with a few of my DIY house projects that the right tool is critical. For example, if I want to turn an Allen-head nut, I can't use a screwdriver. To clarify the tools I'm using for the tasks I've set out in this book, I need to define some key terms. With the word *faith* I am highlighting a more personal side of religion. *Theology* is a generic term for a set of beliefs and practices that associate people with God, gods, or ultimate reality. *Christianity*, or the "Christian religion" (although we don't use that term often), is what we believe and do in light of God's revelation in Jesus Christ through the Holy Spirit. Faith emphasizes the subjective side of religion, and theology is the reasoned reflection on one's faith. To say "I believe" is to set out that you have faith; to say "This is what I believe" is to begin to work on your theology.

All these terms are employed within the Christian tradition, particularly with an evangelical accent. By *evangelical* I mean those followers of Christ who are particularly related to the evangel, the "good news" of the gospel, and thus to salvation in Jesus Christ and to the Bible. I am not using the term in its political sense, which I take to be quite—if not entirely—different. For various reasons, some today prefer the term "Christ-follower," but I'll stick with the garden-variety "Christian" and "Christian faith."

Mere Christianity also represents what theologian Thomas Oden calls the great "consensual tradition"[16]—in other words, the consensus of the body of Christ (which is equivalent to "the church" in this book) over the past two thousand years. It's what Jude 3 describes as the "faith that was once for all entrusted to God's holy people." That's a faith worth fighting to preserve. Mere Christianity is not one party or one flavor but what Vincent of Lerins describes as being believed "by all Christians, in all times, and in all places." Christianity is particularly well-summarized in its trinitarian belief in the Nicene Creed of AD 381 and the Chalcedonian Creed of AD 451, the latter defining Jesus Christ as "truly God and truly man."[17] The best of evangelical Christian faith falls into this great tradition. And as Alasdair MacIntyre has insightfully noted, this tradition is not some homogenous form of opinion; instead it's "sustained and advanced by its own internal arguments and conflicts."[18] Some might notice that this sets evangelicalism within the wider tradition of orthodoxy, which I think is a helpful reminder that those who respond to Jesus Christ as Lord and Savior aren't limited to an evangelical confession.

Let's also take note here of the structure of the Nicene Creed: first it discusses the Father (first article), then the Son (second article), then the Holy Spirit (third article). Theologians ever since have organized their doctrines around the three persons of the Godhead. Around the Father they discuss creation, the fall, teachings about humankind (or anthropology), and so on. You'll notice that most of this book—as has been true for science and faith generally—will circle around "first article" topics. But from time to time we'll also tackle second- and third-article themes.

I am convinced that we can affirm all these stirring, robust confessions of faith along with the findings of modern science. That's why I'm advocating for Christians to engage mainstream science. But what is *science*? Synthesizing various definitions, I define science as "study of and knowledge about the natural world derived through observation and experiment." It's worth noting that *science*, from *scientia* in Latin, simply meant "knowledge" until it was applied more specifically to the natural sciences in the nineteenth century. The related term *technology* (which we'll discuss in chapters six and seven) is "the use of science in industry, engineering, etc. to invent useful things or to solve problems" and "a machine, piece of equipment, method, etc., that is created by the use of science."[19] As we'll discuss later, science and technology are becoming much closer in the minds of emerging adults.

All this adds up to the conviction that, before we seek to integrate science and faith, we have to grasp their inherent differences. Theology at its core focuses on God, who is supernatural—that is, beyond or above nature (*super* means "above" in Latin, as in "not defined or limited by"). Science on the other hand is limited in scope to the natural world and its interactions and laws. This is the meaning of *methodological naturalism* (an often misunderstood term, particularly in the Christian world): the methods of science are designed to find what the natural causes are.

I hope these definitions and clarifications will offer some road signs as we journey forward.

Manifesto and Field Guide

Before I conclude, there's a little more to say about the topics I'm addressing in this book. First of all, we who care about emerging adults and their spiritual lives have to understand how they're changing the landscape of science and faith. We have to take on the early chapters of Genesis and related biblical texts and grapple with what they mean about the age of the earth, human evolution, Adam and Eve, and generally how God created us and this world. But emerging adults care about more than pure science. So I will take us on a tour of technology, its positives and negatives, and also make side trips into New Atheism, cognitive science, the Big Bang, cosmic fine tuning, Intelligent Design, sex, science, and global climate change. I'll conclude with critical strategies for moving forward.

Ultimately, I'm arguing for why we as a Christian community need to bring together Christianity with mainstream science, what that looks like, and how to do this task.

This book is aimed primarily at pastors and emerging-adult ministry leaders, as well as eighteen- to thirty-year-olds who take science and Scripture seriously. It should serve as both a manifesto and a field guide. As a manifesto, it's designed to convince you that the church must embrace mainstream science for its future. My hope is that you'll do something as a result of reading the pages that follow—that you'll write some new, true narratives that integrate Scripture and science and that will speak to emerging adults. Some of those narratives will develop from the eighteen- to thirty-year-olds' lives you'll influence. As a field guide, this book presents a picture of what it looks like to pursue this kind of work—the challenging mountains, the gorgeous vistas, the dangerous sinkholes, and the peaceful meadows. Most of all, I hope you find a true Companion on that journey, perhaps similar to the disciples who trekked to Emmaus and found that Jesus had been walking with them as an unknown traveler (see Luke 24:13-32).

I close with one final experience. I've taught about faith and science in churches for the past two decades, always with an eye to what it means for college and postcollege emerging adults. In a research project I headed up, Science for Students and Emerging, Young Adults (SEYA), our team looked at how eighteen- to thirty-year-olds' attitudes on faith and science form and change. We had target groups in New York City, Menlo Park, and my current homestead of Chico, California. One of our key findings was that even if many emerging adults perceive that the teachings of science and religion conflict, when someone they trust discusses the topic and demonstrates integration—a pastor, college group leader, or friend—they want to hear more.

Sarah, a grad student at UC Davis in a science-related field, drove one hundred miles to attend a workshop at my church in Chico on integrating Genesis 1 and 2 with the Big Bang, quantum cosmology, and evolutionary science. I finished the talk with insights from Tim Keller, C. S. Lewis, and John Stott, which led me to close with this: "So you can see—that's why a robust commitment to Scripture can be brought together with the best of modern science. It's exciting!" Sarah immediately came up afterward with similar enthusiasm. "I loved this stuff! Why don't we hear more about this in church?"

To answer that question is the goal of this book. Let's start our expedition by surveying the landscape of emerging adult culture.

2

EMERGING ADULT FAITH

NOT AN LP, BUT A DIGITAL DOWNLOAD

[Being a spiritual bricoleur] involves piecing together ideas about spirituality from many sources, especially through conversation with one's friends. We have seen that spiritual choices are not limited to the kinds of denominational switching that some scholars are content to emphasize. . . . It also takes the form of searching for answers to the perennial existential questions in venues that go beyond religious traditions.

ROBERT WUTHNOW

---

here is the interaction of science and religion headed? Is there going to be a fight to the death? A pleasant détente? Or something collaborative?

Apparently this question was on the mind of mathematician and philosopher Alfred North Whitehead when he was nearing retirement in the mid-1920s. One of the greatest intellectuals of the late nineteenth and early twentieth centuries, Whitehead had coauthored the epic tome *Principia Mathematica* with the celebrated philosopher Bertrand Russell on the philosophy of mathematics while at Cambridge. He had also labored to develop the mathematics that undergird relativity theory. He was continually dedicated both to the practice of and reflection on science. Robert Lowell, then president of Harvard University, had an idea. Knowing this intellectual star was finishing his time at the Imperial College of Science and Technology in London and that English law was about to force his retirement, Lowell invited the sixty-three-year-old Whitehead to Harvard in 1924 with a purpose: to set aside his purely scientific pursuits for sustained reflection on a new philosophy of nature. When Whitehead's wife asked him about the offer, Whitehead replied, "I would rather do that than anything in the world."[1]

In one of Whitehead's first books during those early years at Harvard, *Science and the Modern World* (1925), he reflected on the way science is embedded in Western culture:

> When we consider what religion is for mankind and what science is, it is no exaggeration to say that the future course of history depends upon the decision of this generation as to the relations between them. We have here the two strongest forces (apart from the mere impulse of various senses) which influence men, and they seem to be set one against the other—the force of our religious intuitions, and the force of our impulse to accurate observation and logical deduction.[2]

We are far from his generation, but "Whitehead's Challenge" still has merit.

In a somewhat different location on the intellectual food chain, the quirky independent film *Nacho Libre* provides commentary. In the movie, Nacho (or Ignacio) cooks for a Mexican monastery but moonlights as a tag-team wrestler, a "luchador libre," with his partner, Stephen. Nacho is a man of faith, but Stephen denies any belief in God and declares that he is "a man of science." This is a point of contention and it makes Nacho feel fearful, especially just before they fight the team of "Satan's Cavemen." In the dressing room before the fight, while Stephen is looking another direction, Nacho surprises Stephen by ambushing him from behind with a guerilla baptism, dunking his head in a plastic basin. My point (yes, I do have one) is that in church we often baptize scientists and the insights that seem convenient to our theological purposes while they're not looking. We're afraid to enter the nitty-gritty of really understanding what scientists do and think. I'm convinced that this is not the best course for the future.

> "I believe religion is holding us back as a species."
>
> *DEVAN, CHICO STATE UNDERGRADUATE*

Sketching Emerging Adulthood

Whitehead's questions—and Nacho's answers—have rattled around in my brain for more than three decades. I already mentioned the Haussauer Problematic, the challenge laid down by my comparative literature professor. When I started pursuing church ministry, that problematic continued to trouble me. In my first year at Princeton Seminary, Diogenes Allen started his lectures on philosophy by addressing the challenges brought to Christian belief by Hume, Kant, and scientific thinking. Ultimately, he told us, these two Enlightenment titans and their concerns wouldn't prove insurmountable, but the point still stuck: scientific rationality posed numerous challenges to the Christian faith. I then traveled to Germany and studied how Whitehead's philosophy interacted with Karl Barth's theology, which culminated in a PhD from the Graduate Theological Union and its Center for Theology and the Natural Sciences.

During my church ministry years I tried to bring all of this study to the congregations I served, which were filled with college students (First Presbyterian Church, Berkeley), postcollege twenty-something urban professionals (Fifth Avenue Presbyterian, New York City), and more college students (Bidwell Presbyterian, Chico). Now I teach at Chico State University. My church and academic years have been filled with emerging adults, faith, and science.

What have I learned? Considering the Haussauer Problematic, Whitehead's Challenge, and Nacho's Answer, how are emerging adults changing the conversation of faith and science? Who, in fact, are emerging adults? What are their attitudes about the changing world they are both inhabiting and creating? And where is the integration of science and faith headed in the future? To answer these questions, we'll look at the work of Princeton sociologist Robert Wuthnow, along with that of Jonathan Hill, Elaine Howard Ecklund, and Christian Smith.[3] We'll also consider insights gathered through my team's eighteen-month SEYA research on the attitudes of eighteen- to thirty-year-olds toward faith and science.[4]

Let me say a few brief words about how this study was conducted. A SEYA team first met with emerging adult participants to teach about the integration of science and religion and hold informal discussions. We presented groups in Northern California and in New York City—a total of 638 participants—with a questionnaire based on surveys from Smith's *Souls in Transition* and Kinnaman's *You Lost Me* and asked these groups to study and to discuss key resources.[5] This usually happened during a four- to six-week period, and we surveyed them before and after to discern whether experiencing this religion and science curriculum made a difference in their attitudes about the possibility of integrating the two.

I also convened a group of twelve thought leaders in emerging adult culture—biblical scholars, theologians, pastors, and a seminary president and dean—to discuss their insights. Finally, I conducted (and am continuing to conduct) in-depth qualitative interviews with thirty college students and postcollege emerging adults.[6] (These will appear throughout the chapters.) These interviews offer contour and nuance to cold statistics because attitudes on religion and science don't fit neatly into four or five boxes, only one of which you can check. Interestingly, my analysis has mirrored the conclusions of wider demographic research from leading scholars.

Feeling "In-Between"

The term *emerging adulthood* may not be entirely familiar. Since I'm analyzing eighteen- to thirty-year-olds primarily from the angle of psychological developmental, it's appropriate to note that a psychologist, Jeffrey Arnett, first defined this category in 2000 as a stage of life in which a person no longer feels like an adolescent but is not yet fully an adult.[7] Though I have some questions (for example, how well it applies universally across the United States), I find the paradigm illuminating. It recognizes the current cultural shift in which individuals are reaching the five milestones of adulthood—leaving home, finishing school, becoming financially independent, getting married, and having children—later than they did in the past.[8] A 2009 analysis found that in 1960, two-thirds of young adults had achieved all five of these markers by age thirty, but by 2000, less than fifty percent of women and one-third of men had done so.[9] We could simplify this by focusing (with Wuthnow) on two markers: Americans are marrying and having children later.[10]

These factual markers have psychological importance. Many of us used to know where to go when we turned eighteen—we graduated from high school, left home and headed to college, landed a job, got married, bought a house, and had kids. For many this path felt stultifying, but it was at least clear. Today, an increasing number of eighteen- to thirty-year-olds feel "in between." Arnett portrays emerging adulthood using five interrelated characteristics: (1) Emerging adults are actively looking for personal meaning and identity. (2) Their lives are marked by instability due to regular relocations, job changes, and revision of life plans. (3) They tend to be self-focused, liberated from parental oversight and significant responsibility for others. (4) They feel (here's the summary phrase) "in between"—beyond adolescence but not yet full adult. (5) Finally, they live in an "age of possibilities," optimistic about the future and keeping their options open.[11] Naturally this is a generalization; those at lower socioeconomic levels often develop job skills more quickly and thus move out of adolescent life and into adult responsibilities earlier than their peers of greater affluence.[12] With that caveat in mind, it's still worth remembering that eighteen- to thirty-year-olds are often markedly uncommitted to church life and marriage. Since the marrying age is around twenty-eight for men and twenty-six for women,[13] most emerging adults' relationship to faith is not defined by family. Their reality presents a jarring contrast to the organization of most congregational ministries. Simply put, most churches are focused on the family (as it were) and have no place for emerging adults in their twenties. As Arnett puts it, "Having left the dependency of childhood and adolescence, and having not yet entered the enduring responsibilities that are normative to adulthood, emerging adults often explore a variety of possible life directions in love, work, and worldviews."[14]

As one who became a Christian at the cusp of emerging adulthood and led college and postcollege ministries for two decades, I read these characteristics with empathy, excitement, and concern. I'm expectant to see how emerging adults change the way we receive and live out the gospel. The future isn't what it used to be (to paraphrase the late prophet and Yankees catcher Yogi Berra). Members of this generation are making great contributions through their creativity, energy, and openness to new ideas. They are, after all, continuing to define what it means to live in a "digital age" (a topic to which we will return). Over my years with this demographic, I have come to treasure their spontaneity and believe that they will make a difference despite enormous challenges such as rising debt, overpopulation, and global climate change.

I also wonder how we who are older have prepared this generation. Frankly, I think we have often failed, as evidenced by the deep anxiety associated with their "in-betweenness." Here I can speak from experience—I've been the pastor of hundreds of emerging adults. I've held numerous conversations in my church office or at Peet's Coffee, talked with them while building houses in Mexico, chatted casually at Christmas parties or after small group meetings. Consequently, this statement from emerging-adult ministry specialists David Setran and Chris Kiesling brings me pain:

> Because many of the stable and scripted road maps of the adult life course have vanished, there is little clear direction on how to proceed through the twenties. In a period of instability, continual change, and new freedom, the weight of personal responsibility can be overwhelming.[15]

As a result of this instability, some even talk of a "quarterlife crisis."[16] In other words, the challenges of twenty-something life may be more extreme than those of midlife.

Just in case you haven't become fully despondent, psychologist and counselor Meg Jay argues in *The Defining Decade* that the twenties define the rest of one's life, particularly in the areas of work, love, and how our brains relate to our bodies.[17] Note that Jay doesn't espouse Christian faith as a resource for the problems she presents—suffice it to say that her chapter on marriage begins with a call for more teaching on relationships for twenty-somethings and continues through a description of the "cohabitation effect" (the fact that cohabiting before marriage is correlated with higher rates of divorce). In accordance, I find that emerging adults are adrift in their romantic (and often sexual) relationships as a result of their generally being undirected.

I mentioned earlier a meeting of twelve of the best thought leaders in emerging-adult culture I could find. As I sat with these Christian leaders in a Fuller Seminary conference room with southern California sunlight filtering through the windows (and through the smog), they described this demographic with some important words and phrases. One observation was that the lives emerging adults live are "marked by instability in relationships, purpose, and faith." Similarly, we found that the word *anxiety* resonated strongly with our combined years of young-adult pastoral ministry. As Setran and Kiesling conclude, "While many of these emerging adult changes can be exhilarating, they also tend to produce a great deal of anxiety."[18]

It is this openness and related anxious stress that characterize the lives of eighteen- to thirty-year-olds and set the context for their culture. Moreover, these specialists commented, emerging adults "are caught in the pastiche of postmodernism. At times they can display delayed self-awareness." They added somewhat harshly if also memorably that eighteen- to thirty-year-olds can be "choice-phobic." Choice phobia obviously relates to relationships but can also characterize commitment to following Christ and to Christian community. This generation has been nurtured in tolerance as a response to pluralism. Nonetheless—lest I sound like this session was one huge complaint session—these leaders all affirmed their love for emerging adults and their excitement about the possibilities they bring to the church, including their incredible technological savvy and ability to take in a wide range of ideas.

So is this new reality good or bad? Smith, who tends toward the negative in his assessment, still summarizes well both the positive and negative sides of the emerging-adult experience:

> The features marking this stage are intense identity exploration, instability, a focus on self, feeling in limbo or in transition or in between, and a sense of possibilities, opportunities, and

unparalleled hope. These, of course, are also often accompanied . . . by large doses of transience, confusion, anxiety, self-obsession, melodrama, conflict, disappointment, and sometimes emotional devastation.[19]

It's worth noting that Smith's follow-up to his first study highlights the shadow side of emerging adulthood, as the subtitle makes clear: *Lost in Transition: The Dark Side of Emerging Adulthood.*[20] Not all is right in Denmark—or at least with emerging adulthood.

It is, of course, entirely possible and utterly faithful for emerging adults to transform their experience of being "in between," with its consequent worry, into a radical openness to what God can do. I've seen plenty of eighteen- to thirty-year-olds do just that. In that light, Eugene Peterson's paraphrase of Philippians 4:6-7 is brilliant:

> Don't fret or worry. Instead of worrying, pray. Let petitions and praises shape your worries into prayers, letting God know your concerns. Before you know it, a sense of God's wholeness, everything coming together for good, will come and settle you down. It's wonderful what happens when Christ displaces worry at the center of your life. (*The Message*)

Notice the displacement of worry with Christ. That's a powerful image. I'm hoping this generation will take the raw material of emerging adulthood, center it on Christ, and let God do a new thing (Isaiah 43:19) in all kinds of areas, including science and faith.

LP or Digital Download? Three Reflections

This sketch of emerging adulthood and faith leads me to an interim question: What would it look like for Christian ministries to engage science and faith in ways that resonate with eighteen- to thirty-year-olds? Here are three possibilities.

First of all, we'd take on some new topics. Technology—for example, the psychological and behavioral effects of screen time, the possibility of artificial intelligence, the promise of transhumanism—has become central to science and religion in the past decade or so; it's not just pure science anymore (Big Bang cosmology and evolution, step aside). In addition, we'd look at new concerns about sexuality and gender (not something you'd find in a standard religion and science textbook). And finally, we'd examine the materialistic philosophy often enshrined in "the findings of neuroscience" that there is no immaterial soul. We'll discuss these in more detail later on.

Second, we'd understand faith in a new way. If any one thing characterizes contemporary emerging adult life, it's pluralism—this generation has been formed in an age of dazzling diversity of all kinds, including that of worldview, religion, sexual identity, race, and ethnicity. Life is open, holding more possibilities than it ever has for past generations. In a worldview or religious context, this means that emerging adults often live, in varying degrees, as "spiritual bricoleurs." And if "bricoleur" doesn't work for you, try "tinkerer"—it means the same thing.[21] This open spiritually means emerging adults often are uncommitted to religious institutions in general and churches in particular. All of this leads Smith to conclude that "emerging adults are, on most sociological measures, the least religious adults in the United States today."[22] This gives many congregations pause and, frankly, prevents them from seeing ministry with eighteen- to thirty-year-olds as a worthwhile investment (not to mention that these young people don't often contribute much to the collection plate—partly because they don't carry cash or checks).

"When it comes to religion and spirituality, I think I would describe myself as more religious than spiritual. I attend mass occasionally and spend time praying when I'm troubled or thankful."

MARIA, AGE 18

Third, we'd see Christian faith as a Spotify mix instead of a vinyl LP, with listeners choosing from a variety of artists based on a certain mood or a feel. At this point, I sometimes tell my college students that it's time to take a trip down memory lane. In the old days I used to drive to Tower Records and thumb through aisles of twelve-inch-square card-board album covers, looking for the right "long-play" recording. Afterward, an LP or two tucked under my arm, I'd journey home, take musical refuge in my room, place the vinyl LP on a turntable, and listen to side A in the sequence the band had set—not an order I constructed. Then I'd go to side B. One album. One sequence. Today, music isn't bound by the sequence set by the artists. Neither is our belief. We have the ability to mix music based on the algorithmic features of Pandora or our iPhone. The listener, not the musician, determines the sequence of the music today. The parallels with faith are reasonably clear.

This means that emerging adults are leading us out of the two-dimensional "science and religion" dichotomy to something much more multidimensional. Here's what this looks like in my ministries: a college student might attend worship at Bidwell Presbyterian Church, take part in a weekly InterVarsity meeting, participate in a small group put together by Cru, go on a yearly mission trip to Mexico with another church, then do a little yoga on the side without any sense of concern. He didn't take the Bidwell Presbyterian LP off the shelf and play through our selected sequence of worship service–small group–large group–mission trip. Emerging adults are experimenting with various spiritual inputs both inside and outside congregations and campus ministry groups.

Put a slightly different way, faith and science are becoming increasingly pluralistic. Talking about "religion and science" may seem like a dialogue between two things, but eighteen- to thirty-year-olds see it differently. They have grown up in environments saturated with options and possibilities. And though pluralism is certainly not new—the apostle Paul referred to the many "gods" in Corinth, and the United States is known as a "melting pot," for example— the experience of it has increased through the explosion of knowledge on the Internet. Consider that the number of websites has topped one billion.[23]

> "You can be a Christian, but it never hurts to meditate."
>
> LUPE, AGE 19

Admittedly, it's also a question whether science is one thing or whether there are multiple sciences, and this further complicates the interaction. In many languages—French, for example—the term is usually plural. So to be a science major in school is literally "to study the sciences." Similarly, British English speaks of "maths" in the plural. Even if we make it a singular word, there are many kinds of math that correspond to different areas of study.[24] Similarly, if we think of "science" in the singular, it is at minimum a set of disciplines. Biology and chemistry are not the same areas of study with precisely the same methods.

So can the church make the engagement of mere Christianity with mainstream science a Spotify mix? I'm not suggesting that taking a fresh approach to science and faith will halt the exodus of the nones and emerging adults will suddenly flood back into churches. But I think the climate has changed. And that is a fact we can't deny.

Speaking of the "nones," one popular portrayal is not very pluralistic at all. A loud cultural narrative is increasing loss of belief as knowledge and science devour faith. Science wins. Faith loses. Let's look at a case study about the "New Atheists," who are largely responsible for this narrative.

ADDRESSING
THE NEW ATHEISM

Question: What were the greatest influences in coming to your decisions about religion and science? Answer: "Watching documentaries about Richard Dawkins and Neil Degrasse Tyson. And friends— a discussion of what they've seen on the Internet."

TREVOR, FIRST-YEAR UNDERGRAD

S cientifically informed atheism affects emerging adults today. That makes it worthy of a brief consideration and response.

The New Atheism and its "four horsemen"—by which I mean Richard Dawkins, Christopher Hitchens, Daniel Dennett, and Sam Harris—arose in the 2000s after the horrors of religiously informed violence. This leads some people to conclude that the wave of New Atheism has already crested—it was brought on by the religiously motivated violence of 9/11 and is now receding into the past.

I'm not entirely sure that's true. With ISIS and other militant groups making news regularly, there's still plenty of violence to fuel slogans such as "If you really want violence, try religion." I also still hear the voices of the New Atheists echoed by my students and other emerging adults. Besides that, materialism (a worldview focused exclusively on the material world) has been around for millennia. So whether this atheism is "new" is in question—it's a recurrent theme of humanity that also affects today's emerging adults.

I should mention that I don't often encounter the sneer and hard-edged approach to Christian faith that Dawkins emits and embodies. The key value I find in eighteen- to thirty-year-olds is tolerance. The attitude of atheist students today is kinder and gentler than that of my Berkeley classmates in the eighties, who seemed intent on disproving my faith. In addition, it's important to note that these New Atheists don't speak for all scientists. Consider what one nonreligious scientist had to say about Dawkins: "He's much too strong about the way he denies religion. . . . As a scientist, you've got to be very open, and I'm open to people's belief in religion. . . . I don't think we're in a position to deny anything unless it's something which is within the scope of science to deny."[1]

So their tone is off, but the New Atheists still have influence, and their ideas are worth our attention. Consider the criticism laid before believers by the late brilliant and voluble Christopher Hitchens in his "Hitchens's Razor": "That which can be asserted without evidence can be dismissed without evidence."[2] In other words, faith, according to Hitchens and his ilk, is not rational. Here's Dawkins again:

"I think most people are neutral—'It's cool if you believe it.' If they want to believe, more power to 'em."

AMANDA, AGE 19, ATHEIST

> What, after all, is faith? It is a state of mind that leads people to believe something—it doesn't matter what—in the total absence of supporting evidence. If there were good supporting evidence, faith would be super-fluous, for the evidence would compel us to believe it anyway.[3]

Oxford theologian Alister McGrath responds, "Dawkins's definition of faith . . . bears no recognizable resemblance to what Christians believe."[4] Accordingly, when presenting the nature of faith in *Mere Christianity,* Lewis wrote, "Faith, in the sense in which I am here using the word, is the art of holding on to things your reason has once accepted, in spite of your changing moods."[5] That sounds jarringly different from the simplifications and canards the atheists are selling—and much more robust and interesting. Overall, the New Atheists have made a name for themselves—it is capitalized after all!—by setting up a number of straw men and then congratulating themselves on their superior intellect when they find them easy to obliterate.

And yet looming around the conviction that the church needs to integrate mainstream science are voices such as Dawkins's saying that evolution makes this impossible:

> I think the evangelical Christians have really sort of got it right in a way, in seeing evolution as the enemy. Whereas the more, what shall we say, sophisticated theologians are quite happy to live with evolution, I think they are deluded. I think the evangelicals have got it right, in that there is a deep incompatibility between evolution and Christianity.[6]

I find it deeply troubling that we choose to get our theology—or even our understanding of evolution—from Richard Dawkins.

It's important to reiterate that not all "nones" are equal atheists. "Nones" are about 30 to 34 percent of the emerging adult population, but atheists represent only about 6 percent.[7] Still, the New Atheists speak for many who are disaffected from the church. That's why Dawkins is so popular among "half-convinced" atheists. It's not his arguments. It's that he says things they agree with and that sound convincing.[8]

Here's how I counter atheism. When Dawkins asserts that some religious people hold beliefs that are childish and irrational, this does not prove that all believers do, nor does it prove that Christian faith is an unreasonable or unreasoned delusion. I might just as readily conclude that all atheists are caustic and arrogant because Dawkins is. But that would be an unjust judgment and untrue.

On Richard Dawkins: "He has brought up many points that I question or have questioned myself, and for this I would say I look up to Dawkins."

DANIELLE, AGE 19

Of course, it's often not rationality per se—especially when it comes to science—that turns people away from faith. It can be an intimate group of friends (the "tribe"), the obnoxious acquaintance who's a Jesus freak, or some personal pain they've experienced that they can't reconcile with the existence of a good God. All of that gets mixed in with what's deemed "rational."

We could turn the tables and point to the historical connection between Christian doctrine and the rise of modern science. Is there any good reason to believe that this universe is coherent and consistent and rational if there isn't a God? Historically, the growth of scientific discovery has been based on the underlying assumption of a rational universe. As the Nobel Laureate and UC Berkeley physicist Charles Townes notes, "For successful science of the type we know, we must have faith that the universe is governed by reliable laws and, further, that these laws can be discovered by human inquiry."[9] Whitehead is good here too: "There can be no living science unless there is a widespread conviction in the existence of an Order of Things, and, in particular, of an Order of Nature."[10] To have a cosmos and not a chaos requires an Orderer. Without that, the atheists' assumption of a rational cosmos is based on faith, not a reasonable deduction from their metaphysics. Science and rationality have an inherent link with God the Creator.

God's ordering our cosmos brings with it the existence of goodness—and the related realities of meaning, purpose, and beauty—and these together present an almost insoluble problem for the atheist.

Thankfully, I didn't actually have to formulate the problem of good on my own. Prominent atheists have already taken on that task for me. Consider Dawkins's words:

> In a universe of blind physical forces and genetic replication, some people are going to get hurt, other people are going to get lucky, and you won't find any rhyme or reason in it, nor any justice. The universe that we observe has precisely the properties we should expect if there is, at bottom, no design, no purpose, no evil and no good, nothing but blind, pitiless indifference.[11]

That's a reasonably bleak portrayal of the universe and, since we're part of that universe, of our lives as well. It does, however, correspond perfectly with a basic tenet of Philosophy 101: nothing comes from nothing. Start with a purely physical system without any Creator, and all you have is brute fact. If the universe is simply a physical system, then why should something nonphysical like goodness, meaning, purpose, or beauty arise? It cannot.

"Science as a worldview is quite popular. 'This is what you can really trust. This you can prove.'"

BRIAN, AGE 23

Margaret Geller, astrophysicist at the Harvard-Smithsonian Center for Astrophysics, concludes that it's pointless to mention purpose in the universe: "Why should it have a point? What point? It's just a physical system, what point is there?"[12] And given certain implied metaphysics—that it's nothing but a physical system—she's right. There will no be point and no good without something from outside infusing the system with these qualities. The words

I'm typing right now have no meaning, no potential goodness or beauty, without the intention of my brain (such as it is), as well as the context that words lend to sentences and thoughts. In the Western cultural tradition, of course, meaning and goodness come not from something but from Someone—a Creator or, better, an Informer who imbues the physical system with nonphysical qualities of information in the act of creation.

But, of course, Western culture is gradually moving away from belief in creation by a good, meaningful, purposeful, and beautiful Creator. Weinberg, the atheistic scientist, states this problem pointedly: "The more the universe seems comprehensible," he asserts, "the more it also seems pointless."[13] Put another way, as we increase in scientific knowledge, we decrease in our ability to comprehend meaning. Pointless indeed. This is the sterility of the narrative of purely materialist science.

Why is this not a greater debate with the so-called New Atheists? More often we hear the challenge, "How can you believe in a good, almighty God when there's so much suffering in the world?" Put simply, atheism confronts believers with the problem of evil.

Yes, there indeed exists an intractable problem of evil. The Bible laid down the quandary millennia ago:

> Mortals, born of woman,
> > are of few days and full of trouble.
> They spring up like flowers and wither away;
> > like fleeting shadows, they do not endure. (Job 14:1-2)

Pain hurts, and it's hard to square with God's existence. For this reason the problem of evil remains the greatest argument against belief in God, and it's been addressed by the great medieval philosopher Thomas Aquinas as well as contemporary thinkers such as C. S. Lewis and Alvin Plantinga.

Still, I offer a counter-question: If you show me a world without a Creator, then you also need to provide grounds for the existence of goodness in this world. How do you intend to answer the problem of good in a world without God?

Often the response is a bald assertion that human beings supply the qualities lacking in a world devoid of God. Back to Weinberg, who comments a bit bleakly on how to find meaning through striving to grasp reality: "The effort to understand the universe is one of the very few things that lifts human life a little above the level of farce, and gives it some grace of tragedy."[14] When asked later in an interview about this comment, he added, "There is a point that we can give the universe by the way we live, by loving each other, by discovering things about nature, by creating works of art. . . . Faced with this unloving, impersonal universe, we make a little island of warmth and love, and science and art, for ourselves—that's not an entirely despicable a role for us to play."[15]

> Question: What's the best relationship between religion and science? Answer: "If it works for people, let it be."
>
> DANIELLE, AGE 19, ATHEIST

I find myself sympathetic to his conviction, but I also hear the thud of an unfounded leap of blind faith. And so, from the mouths of atheists, it sounds like the problem of good is insoluble. And I believe it is—without an act of faith. The question is this: Will this faith be reasonable or not?

In conclusion, let's listen to atheists as voices for emerging-adult concerns, but let's not be overwhelmed by their assertions of rationality. Neither unbelief nor belief is purely intellectual, and sound reasons exist for doubting atheism. Rational argument ultimately doesn't change a person, but sometimes it leads them forward. I know it did for me as a functional atheist meandering down UC Berkeley's Sproul Plaza.

3

EMERGING ADULTS

ARE THEY NONE AND DONE?

I, having been subjected to a Christian baptism before reaching an age of consent, or having submitted to baptism before embracing freethought and reason, hereby officially renounce that primitive rite and the Church that imposed it. . . . From this day forward, I wish to be excluded from any claims of religious affiliation or membership based on baptismal records.

**FREEDOM FROM RELIGION FOUNDATION'S
DEBAPTISMAL CERTIFICATE**

Dan Barker was once a Christian evangelist who also wrote worship music. He describes how this path began with his parents:

My dad was a professional musician during the 1940s. At one of his concerts he met a female vocalist and, as things go, they went (lucky for me). They got married and, when I was a toddler, they both found true religion. Dad threw away his collection of original Glenn Miller recordings (ouch!) [and] turned his back on his former "sinful" life. . . . For a number of years we formed a family musical team and ministered in many Southern California churches—nothing fantastic—Dad played trombone and preached, Mom sang solos, I played piano, my brothers tooted various instruments, and we all joined in singing those famous gospel harmonies. It was a neat experience for us kids. My childhood was filled with love, fun, and purpose. I felt truly fortunate to have been born into the truth and at the age of fifteen I committed myself to a lifetime of Christian ministry.

He continues, "I went to a Christian college, majored in Religion/ Philosophy, became ordained, and served in a pastoral capacity in three California churches. I personally led many people to Jesus Christ and encouraged many young people to consider full-time Christian service."

He still receives royalties from his music: "I have written more than a hundred Christian songs that are either published or recorded by various artists, and two of my children's musicals continue to be bestsellers around the world."

Gradually, Barker found that he couldn't believe it anymore. He became not just a none but an evangelist for atheism: "I did not lose my faith; I gave it up purposely. . . . There was no specific turning point for me. I one day just realized that I was no longer a Christian, and a few months later I mustered the nerve to advertise that fact."

And so he now uses this skill to impugn the Christian message: "I was a preacher for many years, and I guess it hasn't all rubbed off. I would wish to influence others who may be struggling like I did—to influence them to have the guts to think."[1]

Barker is the founder of the Freedom from Religion Foundation (FFRF), an organization that promotes a radical separation of church and state. Barker's deconversion and counter-evangelism efforts led to one of my saddest experiences on the Internet—discovering FFRF's debaptism certificate, a document that records one's rejection of their baptism.[2] Barker, a former pastor who doesn't seem enamored with his previous profession, will send you a certificate to that effect.

It's worth adding here that I don't think of myself as an overly sensitive soul or a person given over to despondency in light of how others see our Christian faith. But reading this I felt teary. I despaired of those moments I remembered as a pastor baptizing infants. (I am Presbyterian, after all—and you Anabaptists, work with me here.) I recalled all the best intentions of those parents (including Laura and me) for our children. We wanted to lay hold of God's covenantal grace and promises. We wanted our children to know that we loved them, that our faith was the very best we could offer, and that good Christian community was a blessing. It wasn't some cruel will imposed on them by Big Brother—or Big Parents.

And yet the fact remains that emerging adults are turning away from the church. These include the "nones," who (like me in my youth) were never baptized or had much Christian teaching to reject, and the "dones," who attended youth group, summer camp, worship services, and small groups but who now are finished with Christianity. Though they were loyal and dedicated for years—even as lay leaders—they are disillusioned with church. *Christianity Today* quotes one "done" as saying, "I guess the church just sort of churched the church out of me."[3]

The Classic Science and Religion Typology

With Barker, the "nones," and the "dones" in mind, I arrive here: if there's one major conclusion we can draw from a cursory look at the data, it's that the church has a huge challenge in asking emerging adults to bring science and faith together. Frankly, they don't seem very interested in faith. In fact, one Barna survey found that a third believe the church does more harm than good.[4]

These broad strokes miss far too much that's significant, so a little intellectual work by way of typology is the next step. Let's relate emerging adults'
attitudes toward faith and science by first taking a look at Ian Barbour's four
ways of relating religion and science. Barbour, the late physicist-theologian
from Carleton College, wanted to set out a simple means to grasp what
happens when science and religion meet.[5] In his own life he brought his
physics training together with a Christian faith informed by process
philosophy. The beauty of his scheme (and the reason for its fifty-plus-years
durability) is that it has only four types. Though almost no one thinks this
typology is perfect, and though we who study the field almost universally
admit there are further nuances to be defined,[6] Barbour's four-part
typology possesses remarkable staying power. It's simple and easy to grasp.
With the caveats in mind, we can categorize emerging adults in all four
categories: conflict, independence, dialogue, and integration.

First, conflict asserts that religion and science will never agree. This
perspective is represented by thinkers such as Richard Dawkins, a former
Oxford zoologist, and fundamentalist Christian Ken Ham, president,
CEO, and founder of Answers in Genesis and the Creation Museum in
Kentucky, who holds a bachelor's degree in environmental biology. We
spent some time on Dawkins and the New Atheists in the previous
chapter, discussing their many conflicts with religion; Ham's approach
to mainstream science is the mirror image. Recently I read an Answers
in Genesis article that took issue with my approach and that of the
organizations I'm affiliated with (BioLogos, of which I am an advisory
council member) and projects I've poured myself into (such as Scientists in Congregations). It asserts that "Christians would be foolish to
believe that no dark forces are working behind the scenes to undermine
the gospel. God warns us to 'walk circumspectly' (Ephesians 5:15)."[7]
This approach takes conflict to the cosmic level, tempting me to ask,
"What about dark forces that work behind the scenes to sow unnec-
essary division among those who profess faith in Christ?"

In a more earthly realm, there's also the Nobel Laureate–physicist Stephen Weinberg, who put forward a model of conflict between the two with great precision and clarity: "One of the great achievements of science has been, if not to make it impossible for intelligent people to be religious, then at least to make it possible for them not to be religious. We should not retreat from this accomplishment."[8] Some students— usually those who are adamantly opposed to religious belief—are hard-core adherents to the "warfare thesis," which was promoted in the late nineteenth century in such key texts as Andrew Dickson White's *A History of the Warfare of Science with Theology in Christendom* and John William Draper's *History of the Conflict Between Religion and Science*.[9] The concept of warfare between science and religion has received an onslaught of scholarly critique, but it has remarkable tenacity even with few scholarly works defending it.[10] It remains prominent among its bestselling apologists—such as Dawkins and Jerry Coyne—and thus in popular culture.[11] I continue to labor over the question of why the warfare thesis has continued to thrive despite such academic resistance.

Along those lines, most emerging adults I've talked with aren't themselves convinced that religion and science ought to be quarrelling. Instead, they've heard about the conflict between the two and are interested in finding a response. Although they know that warfare is in the air, they aren't generally given over to it.

The second view, independence, concludes that religion and science are two completely different ways to look at the world and their boundaries should be observed. Stephen Jay Gould is its most well-known proponent, along with theologian Karl Barth (who famously found no need to consult science in formulating a doctrine of creation).[12] It makes its way into statements from prestigious organizations such as the National Academy of Sciences: "Attempts to pit science and religion against each other create controversy where none needs to exist."[13] It offers a way out in a pluralistic world where

"Science
is science, and
religion is religion.
Not much of a
blend between
the two."

STEVE, AGE 21

emerging adults are fatigued by all the decisions they face and don't see an easy way to reconcile all the inputs they receive. My research indicates that students often take the independent approach when they're not really sure what they believe. Why not just leave them separate but equal? Smith was surprised to find that most eighteen- to thirty-year-olds remain remarkably vague in what they believe, whether about God or science or a host of other topics.[14] In addition, this generation has been nurtured in pluralism and one of its most important coping mechanisms, tolerance.

This brings me back to the great Harvard paleontologist Stephen Jay Gould, who created a noteworthy and oft-quoted version of independence: non-overlapping magisterial authority, or NOMA. He comments, "Science and religion do not glower at each other . . . [but] interdigitate in patterns of complex fingering, and at every fractal scale of self-similarity."[15] I love this quote because most who hear it have no idea what it means. Another comment from Gould is much more straightforward:

> To say it for all my colleagues and for the umpteenth millionth time (from college bull sessions to learned treatises): science simply cannot (by its legitimate methods) adjudicate the issue of God's possible superintendence of nature. We neither affirm nor deny it, we simply can't comment on it as scientists.[16]

Gould's general point with NOMA is that we should respect the boundaries of both science and religion, as well as affirm the legitimacy of each. For what it's worth, Richard Westenberg, the music director at my first pastoral post, Fifth Avenue Presbyterian in New York City, knew Gould personally. He told me several times that Gould was a brilliant and kind man (and a passionate New York Yankees fan) who didn't conclude that "thoughtful Christian" was an oxymoron. Maybe Gould's friendship with brilliant church friends like Westenberg made a difference. Maybe NOMA matched his personality. It may be the same for those who are tired of fighting.

Third, some young adults endorse dialogue or integration. Dialogue often involves a respectful discussion of insights from each discipline, an approach seen at a variety of academic gatherings. Almost every academic theologian or scientist who's been at a science and religion conference has done the work of dialogue. We talk. And talk. And talk. Integration, on the other hand, holds that science and religion (including theology) need to make a difference to each other through collaboration; it's exemplified by people such as Francis Collins and Robert J. Russell, a scientist-theologian.

Question: What's the best relationship between religion and science? Answer: "Probably integration . . . making faith deeper by seeing how science works."
BRIAN, AGE 23

While dialogue is what scholars do at conferences, Barbour really wanted us to move toward integration. In my experience, both dialogue and integration are well-represented among emerging adults. When I begin my science and religion class at Chico State, I often have the students form a visual graph of where they find themselves on Barbour's scale by grouping themselves by their preferred type of interaction in the four corners of the classroom. There are few in the conflict corner, a few more in independence, and the most (usually over half the class) in dialogue and integration. (I also have a fifth spot, in the middle of the room, for those who are undecided. To my surprise, few choose this.)

My experience, both in the public university and in congregations and emerging-adult ministry groups, is that many emerging adults see a profound correlation between science and faith. One student used a homey metaphor that stuck: "Religion and science are like peanut butter and jelly—one can't be without the other." Others recommend creatively choosing components from each. I commonly hear the phrase, "Everyone has right to their own beliefs." Not only do these words ring out through secular classrooms but also within church walls.

Nevertheless, collaboration—even independence—might sound a bit too neat and tidy. Of course, at times there needs to be conflict between science and faith. If a scientific power exists, that doesn't mean we have to use it. An *is* doesn't immediately imply an *ought*. The atomic bomb, for example, is not an unmitigated good simply because it exists. In fact, it needs someone with a wider vision to put on the brakes. (More on this in chapter seven.)

Finally, we know that God is not ultimately demonstrable through natural science. God's fingers, as it were, won't poke through a laboratory experiment. In fact, I believe we want to keep a measure of independence between faith and science. As nice as many pastors are, scientists don't need them in the lab sprinkling holy water on their experiments or providing 24/7 spiritual encouragement.

Though I advocate for integration, independence has a place. If God is going to work in the natural world, he will often do so through natural means. (Though obviously a miracle like the resurrection is a direct act of God without natural causes.) One tried-and-true way to understand this is dual causation.[17] We can speak of an event through two means— God's and the world's. For example, when Moses led the Israelites out of Egypt, "Moses stretched out his hand over the sea, and all that night the LORD drove the sea back with a strong east wind and turned it into dry land" (Exodus 14:21). Did God or nature do it? Yes—both are necessary to describe the event. Similarly, consider Psalm 139:13: "For you created my inmost being; you knit me together in my mother's womb." When God knits us together in our mother's womb, that divine work also occurs through natural processes. This is a useful perspective. We should not use science to prove God's existence, and there is no uniquely Christian way to bring water to the boiling point or to map the human genome.

But in this book I am primarily advocating for integration. Why? Our Creator is the "God of truth," a concept we find repeatedly in the Bible (see Psalm 31:5; Isaiah 65:16 KJV). We find God when we seek truth in Scripture, or in science—or in art, or in business, or in any human endeavor. Another way to phrase this is that Jesus is Lord of all. We join voices with the angel in Revelation that Jesus "is Lord of lords and King of kings" (Revelation 17:14). Therefore in any form of human knowledge—which naturally includes the sciences—we are bound to bring our discoveries under the lordship of Christ. This means that no discovery can dictate our theology or ethics, but also that no form of human insight and knowledge is outside of Christ. Put simply, God knows far more about science than Albert Einstein.

> Question:
> What is the proper relationship between science and religion?
> Answer: "They are independent, and as a sub-effect, they conflict with each other."
>
> *MATTHEW, AGE 18*

In sum: *Seek integration when possible. Address conflict when required. Allow for independence when necessary.* I will thread this rubric throughout the book. Ultimately, the reality often lies somewhere among these three.

Back then to our eighteen- to thirty-year-olds and Barbour's scheme. I've already noted where emerging adults fit, but the types can only set out the general contours. Spiritual bricoleurism complicates the typology. Even more significantly, emerging adults perceive that there is conflict between science and religion in the wider culture, but they generally do not subscribe to this way of relating religion and science. Emerging adults hear about conflict, but they seek collaboration or independence.

To explain how I arrived as this conclusion, I need to move to a tale of two studies.

A Tale of Two Studies

Can science and faith collaborate or do they most often conflict? Depending on where you look, there are some challenging statistics.

In an important study conducted by the influential sociologists Christian Smith and Kyle Longest, 70 percent of 2,381 undergraduate eighteen- to twenty-three years old stated that they "agreed" or "strongly agreed" with the statement that the teachings of religion and science conflict. And more than half (57 percent) disagreed with the statement "My views on religion have been strengthened by discoveries of science."[18]

So it might seem that the case for the warfare (and thus conflict) model among emerging adults is decided. Christopher Scheitle's work, however, provides some competing data. Analyzing UCLA's Spirituality in Higher Education Survey of 10,810 undergraduates, he arrived a seemingly opposite conclusion:

> The analysis finds that, despite the seeming predominance of a conflict-oriented narrative, the majority of undergraduates do not view the relationship between these two institutions [religion and science] as one of conflict.[19]

He found that 69 percent of those surveyed agreed that independence or collaboration was the best way to relate religion and science. Those who did lean toward conflict were divided between those who sided with religion (17 percent) or with science (14 percent).[20] This mirrors Elaine Howard Ecklund and Jerry Park's survey of 1,646 scientists at twenty-one elite US research universities: "In contrast to public opinion and scholarly discourse, most scientists do not perceive a conflict between science and religion."[21]

How do we make sense of these competing claims? A bit of sleuthing led me to this conclusion: it depends on the question posed. I forged more deeply into the studies and found that Smith and Longest asked about the culture at large, and the UCLA study and Scheitle about personal views.

Specifically, Longest and Smith looked at the range of responses agreeing or disagreeing with this statement: "The teachings of science and religion often ultimately conflict with each other." Scheitle analyzed the response to this statement: "For me, the relationship of science and religion is one of . . . " The first is phrased more generally. It asks about the "teachings of science and religion." The second has a personal tone with the phrase "to me"—in other words, it's about personally held attitudes. Simply put, the majority of emerging adults in these studies sense that there is conflict "out there" in the culture, but they themselves want something better than conflict. They are fatigued by the culture wars.

These findings parallel the results of another study looking at the US adult population as a whole: "The new Pew Research Center findings show that most Americans (59%) say, in general, that science often is in conflict with religion," but "at the same time, however, most adults (68%) say there is no conflict between their personal religious beliefs and science."[22] Conflict is in the air, but not in the hearts of believers.

In light of these studies—in addition to my own experience, discussions with specialists, and survey work—I'm willing to conclude that most emerging adults are generally positive about the compatibility between science and religion. It's more that they have heard others arguing for incompatibility. Perception of conflict is the key. Though they know that warfare is in the air, many emerging adults don't want any part of it. Rather, they're interested in finding ways that religion and science are compatible. This brings my conversation with David to mind:

> The best relationship between religion and science is not conflict, nor independence, but more dialogue. It's important to talk with people and get their advice. You know—good conversation. There shouldn't be any fighting. Be more friendly and open. Less conflict and more dialogue. Give it some thought, chew on it for a while.

Therefore, in David's view, Dawkins is a "turnoff."

Here's an interesting element in Scheitle's analysis—almost 39 percent of business majors chose conflict as the way science and religion best correlated.[23] Similarly, a recent Dartmouth graduate commented in an interview with me that in his experience, those who most strongly resisted faith were finance and business majors. As he phrased it, they were the "closest to hardcore materialism" and "extremely driven." Their motto, as he described it, was "In it to win it." (Incidentally, I have nothing against majoring in business and finance—I've preached for decades that we need more Christians in those arenas—but there may be spiritual land mines in these fields; see Matthew 6:24.)

All this reminds me of Blaise Pascal's insight that our resistance to faith is not only intellectual but stems from our resistance to the personal demands of the Christian message. In philosophical terms, the message of Christ is self-involving. In this case, the desire (or, for Pascal, "lust") for wealth and power hinders us from receiving of the gospel:

> In order to make man happy, it must prove to him that there is a God; that we ought to love Him; that our true happiness is to be in Him, and our sole evil to be separated from Him; it must recognize that we are full of darkness which hinders us from knowing and loving Him; and that thus, as our duties compel us to love God, and our lusts turn us away from Him, we are full of unrighteousness.[24]

We have to keep personal and moral factors in view—it's not simply about convincing science-minded people to see *intellectual* compatibility with faith, although that's important. Ultimately, we're after something much more grand and unifying—where our entire being finds integration and we follow Christ, who told us, "Love the Lord your God with all your heart and with all your soul and with all your mind" (Matthew 22:37). Love unites reason and emotions. This is the Christian maturity that Paul prays for in the recipients of his letter to the Philippians:

So this is my prayer: that your love will flourish and that you will not only love much but well. Learn to love appropriately. You need to use your head and test your feelings so that your love is sincere and intelligent, not sentimental gush. (Philippians 1:9-10, *The Message*)

There is a cosmic level to all this as well, for "in him [Christ] all things hold together" (Colossians 1:17). Christ holds together our heart, soul, and mind and also holds together all we learn from science and from Scripture.

Because of this, I believe we as humans are drawn toward integration. Emerging adults may not be personally negative about the compatibility of science and faith, but they certainly continue to hear others argue for incompatibility. Maybe they've watched Bill Nye and Ken Ham on TV or seen YouTube clips of Dawkins or Ray Comfort—these are nearly ubiquitous. So they know that warfare is in the air, but when asked they want a relationship of collaboration, or at least independence, between science and faith. Speaking of evangelicals specifically, Ecklund has found through a survey of more than ten thousand people that evangelical Christians are remarkably interested in science, and roughly half (48 percent) are convinced that science and religion can work in collaboration.[25] This all corresponds with some of the most consistent responses I received in interviews[26]—take Travis, age nineteen, who told me, "I'm really interested to hear from someone who's thought about these issues."

Modeling Integration

Unfortunately I cannot predict that an integration of mainstream science and mere Christianity will dry up the market for Barker's debaptism certificates. But it might help stem the flow of "nones" and "dones" away from the church. So how does integration take shape in emerging-adult ministries? How do we speak to a demographic context that hears conflict but favors collaboration or independence?

First of all, the church needs to show how it engages mainstream science. Don't teach the controversy; teach the collaboration—that's what emerging adults want. The majority want to move beyond warfare. That's also where good theology leads, toward speaking into the culture that surrounds us. For example, the word *logos*, which was used by John to describe the incarnation—"the [logos] became flesh" (John 1:14)—is a Greek term that means "word," "pattern," and even "the structure of the universe." For the Stoics, the *logos spermatikos* was the generative principle of the cosmos that creates and holds together all things. It's akin to the way we use the phrases "spacetime continuum" and "the laws of nature." It is, to quote New Testament scholar N. T. Wright, "a kind of principle of rationality, lying deep within the whole cosmos and within all human beings."[27] I believe John's Gospel employs this term with this meaning in mind to say, "You've got your ways to relate to all there is, and I can show you another—deepest reality is found in a person, Jesus of Nazareth." We too can demonstrate connections between what science discovers about the Big Bang, about spacetime, and about the laws of nature and relate those to Christ and the witness of the book of Scripture.

Second, we need to engage the Internet. The influence of the Internet on this generation is largely negative, and it inflames the conflict narrative. Here's an example from a Facebook post: "The idea of evolution was proposed as a way to eliminate God; why should Christians think it makes sense to mix God back in with it?"[28] I hope this book offers significant reasons why such points of view are wrong and distorted.

> Question: What's the best way to relate religion and science? Answer: "Ideally, I would integrate evolution alongside creation. But if you're going to learn about the faith side of science, it won't be in school."
>
> **KATHRYN, POSTCOLLEGE EMERGING ADULT**

To be fair, we could just as easily select samples from the scientific atheists. As one blog comment predicted, "The Internet will kill religion." Another opined, "Jesus will soon go the way of Zeus and Osiris." And it's not hard to find Internet memes with comments like "Let me introduce you to my bronze-age sky god" when ridiculing Christianity and Judaism. On the Internet, these comments are "clickbait," provocative snippets that demand our attention by their outrageous or adversarial claims.

This, of course, means that we who see mainstream science and Christian faith as compatible must produce better material on the Internet. For example, the members of GC Science—a ministry that engages faith and science at Grace Chapel, a prominent evangelical congregation in Massachusetts—decided they wanted a presence on Facebook to counter those of prominent atheists. As of the writing of this book, they were on their way to six hundred thousand likes on their page, which dwarfs the reach of most atheist pages. They are working to make their voice heard in one dominant cultural vernacular. (More substantively, they also run conferences regularly that feature their own theological reflections as well as those of MIT and Harvard scientists.) Moreover, it's also not hard to find video of Francis Collins or John Lennox on YouTube, any number of excellent Veritas Forum talks online, or the excellent material on the BioLogos website. The key here is that science-minded Christians must continue to engage the web. Eighteen- to thirty-year-olds want something better than conflict, but they don't always know where to look. Quality resources make a difference.

COGNITIVE SCIENCE AND REASONS NOT TO BELIEVE

The neuroscientific worldview—the idea that the mind is what the brain does—has kicked away one of religion's intuitive supports. Even if you accepted all of the other scientific challenges to religion—the Earth revolving around the sun, the evolution of animals, and so on—the immaterial soul was one thing you could keep within the province of religion. With the advance of neuroscience, that idea has been challenged.

STEPHEN PINKER, IN *ATOMS AND EDEN*

One of the arguments against Christian faith I hear from eighteen- to thirty-year-olds is that the idea of the soul does not make sense anymore in light of science. Let's look at that concept in more detail.

Philosophical materialism, the perspective that the material world is all that exists, is quite ancient, but to many emerging adults it seems fresh and new. Some scholars deploy the insights of neuroscience to debunk the existence of the soul in order to provide us with reasons not to believe.

For most Christians, if there's no soul, there's no real experience with God. And so for the past several years I've been convinced that the field of cognitive science presents new challenges and opportunities for fresh insights in the dialogue between science and religion.

The Harvard experimental cognitive psychologist Stephen Pinker (in the epigraph) hits the target on at least one count.[1] Prominent in many Christian circles is a dualistic concept of the soul—the idea that there is an entirely separate substance within our bodies that makes up our soul, like air inside a tire or, more philosophically, what Gilbert Ryle called the "ghost in the machine." This owes much more to Plato than the Hebrew Bible, which begins by describing God creating the first man, *adam*, from the dust, *adamah*, by breathing his breath into him (see Genesis 2:7). We, like Adam and Eve, are a psychosomatic unity of body and spirit. Engaging cognitive science can help us correct our doctrine of the soul by reminding ourselves of the Scripture, which sees human beings as body-soul, a psychosomatic unity.

God's creation of humankind, or theological anthropology, is a first-article doctrine. Cognitive science reminds us that, on the other side of the biblical narrative (the third article of theology, which emphasizes the work of the Holy Spirit), our full redemption will include the body (see Romans 8:23; 1 Corinthians 15:12-57). We are not complete without that body.[2] Our hope is not for some "platonized eschatology" (the phrase is N. T. Wright's) in which our disembodied essence rises up to a noncorporeal heavenly realm.[3] Instead, Christians believe in the resurrection of the body, where we are remade in the new heavens and earth. There is a beautifully moving earthiness that we as Christians have inherited from the Hebrew Scripture.

> I know that my redeemer lives,
>> and that in the end he will stand on the earth.
> And after my skin has been destroyed,
>> yet in my flesh I will see God. (Job 19:25-26)

We are fully embodied beings, created good by a loving Creator, who doesn't want to destroy us but remake and redeem us.

In sum: Christians who have wandered from the Bible's psychosomatic unity are properly corrected by Pinker and other cognitive researchers. Or we can simply remember that a sharp tap on the head with a hammer will probably interrupt our prayer life.

Justin Barrett, a cognitive scientist at Fuller Theological Seminary, has applied his scientific training to develop the cognitive science of religion (or CSR), "a scientific approach to the study of religion that combines methods and theory from cognitive, developmental and evolutionary psychology with the sorts of questions that animate anthropologists and historians of religion."[4] Barrett employs the findings of the cognitive sciences to argue that evolution has developed human beings so that we are predisposed toward teleology: "Evidence exists that people are prone to see the world as purposeful and intentionally ordered,"[5] which naturally leads to belief in a Creator. This starts as a young age. Barrett found that preschoolers "are inclined to see the world as purposefully designed and tend to see an intelligent, intentional agent behind this natural design."[6]

Similar research suggests that evolutionary pressures—particularly the human tendency toward cooperation, which helps ensure survival—produce a common morality. Barrett notes that "that humans seem to naturally converge upon a common set of intuitions that structure moral thought," such as the idea that "it is wrong to harm a nonconsenting member of one's group."[7] Andrew Newberg and Eugene D'Aquili have studied brain activity during meditation and prayer and also discovered cognitive function that supports belief in God.[8] Some, such as Pinker and cognitive anthropologist Pascal Boyer, take this tendency into the service of atheism and use it to impugn belief in God—that is, we cannot help but believe.[9] Our brains demand it. But this view essentially takes atheism as a given and argues to its conclusion. It begs the question.

Why couldn't the argument point in the opposite direction? As part of God's creation we are created with openness to belief. Why wouldn't God create some sign of his existence in our brains?

> "I don't know what 'it' is. Part of life is living in the mystery."
>
> SEAN, 19, DISCUSSING A SPIRITUAL EXPERIENCE

CSR offers us a new understanding of the natural knowledge of God, or what Calvin called the "sense of the divine,"[10] a sense of the numinous, powerful and brooding. "Where can I go from your Spirit?" cries the psalmist in Psalm 139:7. "Where can I flee from your presence?" Do you know the feeling of being out in a forest at night, knowing that no one is there but feeling something? This experience can frighten us. At other times this sense comforts us. It seems then that the cognitive sciences, while challenging our faith, can also call us to revise our theological concepts. CSR also gives us the ability to discern the mystery expressed in Ecclesiastes 3:11, that God "has also set eternity in the human heart." Since "heart" is often actually the kidneys in the Hebrew Old Testament, could we just as legitimately refer to "eternity in our brains"? At any rate, CSR provides electrifying connections with our faith.

In light of this, let's remember to ask the question, "Does this counter-argument make sense? Or are there atheistic assumptions that have been assumed?" Sometimes simply taking a pause and asking questions can offer surprising resolution.

Now onto another topic of critical importance: the Bible and mainstream science, particularly how we read the book of Scripture in light of the book of nature.

4

ON A CRASH COURSE WITH HERMENEUTICS

I look up to the mountains;
does my strength come from mountains?
No, my strength comes from GOD,
who made heaven, and earth, and mountains.

PSALM 121:1-2, *THE MESSAGE*

If I were going to coach antagonists on how to challenge Christian faith, I'd say, "Camp out on progress. And do so subtly. Sell the idea that science and technology look forward and continually improve. That today we know far more about the universe than in the first century. That the current iPhone is faster, smaller, more exciting, and therefore better than when it first appeared in 2007. On the other hand, present how religious knowledge always looks in the rearview mirror. Then find a way to phrase the question, 'Which would you rather follow—what's new and constantly updating, or what's old and outdated?' And by the way, don't work too hard at drawing out all the conclusions or offering much historical backing."

In my college teaching, as well as in interviews and conversations with emerging adults about science and religion, I repeatedly hear that newer is better. Ancient religious texts seem unscientific, and Christianity relies on the Bible, which is old. How many eighteen- to thirty-year-olds read books written two thousand years ago? One college student I interviewed told me that he studied various sections of Bible and found it "just bizarre."

"Most religious texts have degrading aspects —the Old Testament, there's some pretty gnarly stuff in it."

JIM, AGE 20

Emerging adults often have difficulty with biblical texts because they seem obscure and scientifically dubious. Curiously, Christians through the ages have argued that our interpretation of the Bible needs to be updated and correlated with good science. Some notions that may appear "biblical" were never in the Bible in the first place and need to be jettisoned. When I read accounts of the Galileo trial—which, incidentally was not primarily about "science versus religion" since both Galileo and his adversaries were believing Catholics and much of Galileo's science was disputed[1]—there was considerable disagreement about how to interpret the Bible's assertion that the sun circled the earth. Psalm 19:6 says the sun "rises at one end of the heavens and makes its circuit to the other." In Galileo's day, the cosmology of Ptolemy and Aristotle put the earth at the center of the solar system. Today, I doubt many Christians would conclude that the Bible teaches geocentricism but would agree that the consensus of science changed and revealed a better interpretive approach. The image of the sun running its "circuit" around the sky doesn't teach that it revolves around the earth but metaphorically depicts what we see every day. In the same way Galileo—along with Augustine, more than twelve hundred years before—concluded that if scriptural interpretation disagreed with known science, we should work with our interpretation to be sure we aren't importing defective science.[2]

Hermeneutics

Principally, resolving the problems of interpreting the Bible in an age of science and technology involves a better, more nuanced, and richer approach to the Bible. And this puts us on a crash course with hermeneutics. So what is hermeneutics? The term derives from the Greek *Hermes*, the name of the messenger god. Just as Hermes delivered messages, so hermeneutics is about how words deliver meaning.

In order to respond to the issue of ancient texts and their contemporary relevance, a great deal depends on how we look at the Bible. What "answers" are in Genesis? The sixteenth-century Reformer John Calvin commented in his *Institutes*, "If we regard the Spirit of God as the sole fountain of truth, we shall neither reject the truth itself, nor despise it wherever it shall appear, unless we wish to dishonor the Spirit of God."[3] Let me parse three distinct and critical elements of this sentence. (1) The Spirit is "the sole foundation of truth"—however and whatever we learn (including through the sciences), God the Spirit is ultimately the teacher. (2) "Wherever it shall appear"— Calvin is referring to what he has learned from secular philosophers like Cicero, but we can certainly extend this to include secular scientists, who are neutral or even antagonistic toward faith. And (3) if we don't refuse to learn from these sources, we're not just neutral or unintelligent, but we actually "dishonor the Spirit of God." Calvin, who was trained as a lawyer, is notching up the rhetoric to make his case.

In light of Calvin's admonition, which I take to be worth following, one reason for teaching on science and its relation to Christian faith is that we need to learn natural science and follow its developments. There's an impressively strong consensus of scientists—98 percent of those in the largest scientific organization in the world, the American Association for the Advancement of Science—who are convinced that human beings and living things have evolved over time; only 2 percent conclude that living beings have existed in their present form since the beginning of time.[4] This implies the age of the earth and universe because time is needed for this evolution to occur.

I assume that most readers of this book are committed to the primacy of Scripture, the view that biblical texts are the unique authority, and a large portion subscribe to biblical inerrancy or infallibility. I'll let Wheaton College, one of the leading evangelical academic institutions, define inerrancy: "The Scriptures of the Old and New Testaments are verbally inspired by God and inerrant in the original writing, so that they are fully trustworthy and of supreme and final authority in all they say."[5] The well-known and oft-used 1978 Chicago Statement on Biblical Inerrancy states that "inerrant signifies the quality of being free from all falsehood or mistake and so safeguards the truth that Holy Scripture is entirely true and trustworthy in all its assertions."[6] But not all evangelicals—certainly not all orthodox, confessional Christians—highlight inerrancy; they would rather use infallibility. One of the premier evangelical seminaries in the United States, Fuller (where I'm on the faculty), has put forth this statement: "All the books of the Old and New Testaments, given by divine inspiration, are the written word of God, the only infallible rule of faith and practice."[7]

Worth noting is that there is no statement about biblical inerrancy or infallibility in the early creeds. The historical, creedal consensus of mere Christianity took the Scripture to be authoritative and the final rule for faith and salvation. Likewise, C. S. Lewis made it clear that he disagreed with inerrancy, which he called "Fundamentalist," and didn't concern himself with whether the Bible contained errors or not.[8] He simply believed it was authoritative, read it daily, and sought to live out its message. I've been told that the brilliant conservative biblical scholar F. F. Bruce (whose commentaries consistently guide my study) once quipped, "We have to do better than a double negative" (referring to an emphasis on the Bible's lack of error). In other words, we need to affirm something more significant. As Christians we ought to focus not on what the Bible doesn't do (in this case, err), but on what it does do. What then does the Bible do? It is the ultimate rule of faith and practice, and so it guides us as Christians. It is a "lamp for my feet, a light on my path" (Psalm 119:105).

Key Questions and Biblical Texts on Creation

In that light, let's look at three critical questions: (1) What do we mean by the teaching that God created the world? In other words, what does the Bible teach us about creation? (2) How long are the six days of creation? Are they twenty-four hours like our days?[9] (3) And all in all, what is the Bible designed to do? How do we read it and interpret it? How do we let it be the authoritative word to us from God? What message is the text actually carrying?

Let's keep these questions in mind as we read the central biblical texts on creation (keeping in mind that there are more texts than these).[10] Sometimes it's useful to simply read Scripture and listen to what it says, to pause and take in what we can from it.

Genesis 1. In these selected texts we see the grand vision of God's creation of the universe and image-bearing humans.

> In the beginning God created the heavens and the earth. Now the earth was formless and empty, darkness was over the surface of the deep, and the Spirit of God was hovering over the waters.
>
> And God said, "Let there be light," and there was light. God saw that the light was good, and he separated the light from the darkness. God called the light "day," and the darkness he called "night." And there was evening, and there was morning—the first day. (Genesis 1:1-5)
>
> Then God said, "Let us make mankind in our image, in our likeness, so that they may rule over the fish in the sea and the birds

in the sky, over the livestock and all the wild animals, and over all
the creatures that move along the ground."

So God created mankind in his own image,
in the image of God he created them;
male and female he created them.

God blessed them and said to them, "Be fruitful and increase
in number; fill the earth and subdue it. Rule over the fish in the
sea and the birds in the sky and over every living creature that
moves on the ground."

Then God said, "I give you every seed-bearing plant on the face
of the whole earth and every tree that has fruit with seed in it.
They will be yours for food. And to all the beasts of the earth and
all the birds in the sky and all the creatures that move along the
ground—everything that has the breath of life in it—I give every
green plant for food." And it was so. (Genesis 1:26-30)

Psalm 19. Notice how the first nine verses of this psalm bring together
the truth of God's presence in the natural world with the reality of God's
revelation in Torah. Notice that there's no break between verses six and
seven; the psalm moves directly from God's revelation in nature to God's
revelation in Torah, here meaning "law" but also "instruction."

The heavens declare the glory of God;
the skies proclaim the work of his hands.
Day after day they pour forth speech;
night after night they reveal knowledge.
They have no speech, they use no words;
no sound is heard from them.
Yet their voice goes out into all the earth,
their words to the ends of the world.
In the heavens God has pitched a tent for the sun.

It is like a bridegroom coming out of his chamber,
like a champion rejoicing to run his course.
It rises at one end of the heavens
and makes its circuit to the other;
nothing is deprived of its warmth.

The law of the LORD is perfect,
refreshing the soul.
The statutes of the LORD are trustworthy,
making wise the simple.
The precepts of the LORD are right,
giving joy to the heart.
The commands of the LORD are radiant,
giving light to the eyes.
The fear of the LORD is pure,
enduring forever.
The decrees of the LORD are firm,
and all of them are righteous. (Psalm 19:1-9)

Psalm 104. I like the way Eugene Peterson renders this poetic celebration of creation.

What a wildly wonderful world, GOD!
You made it all, with Wisdom at your side,
made earth overflow with your wonderful creations.
Oh, look—the deep, wide sea,
brimming with fish past counting,
sardines and sharks and salmon.
Ships plow those waters,
and Leviathan, your pet dragon, romps in them.
All the creatures look expectantly to you
to give them their meals on time.

You come, and they gather around;
> you open your hand and they eat from it.
If you turned your back,
> they'd die in a minute—
Take back your Spirit and they die,
> revert to original mud;
Send out your Spirit and they spring to life—
> the whole countryside in bloom and blossom. (Psalm 104:24-30,
The Message)

Isaiah 45:18. In this section of Isaiah, we read some of the clearest statements of God's nature as the Creator.

For this is what the LORD says—
he who created the heavens,
> he is God;
he who fashioned and made the earth,
> he founded it;
he did not create it to be empty,
> but formed it to be inhabited—
he says:
"I am the LORD,
> and there is no other." (Isaiah 45:18)

Proverbs 8. In this passage wisdom is personified and described as having a unique role in creation. It's paralleled in many of the New Testament passages about Christ and creation. I like the way the New Revised Standard Version renders this section.

When there were no depths I was brought forth,
> when there were no springs abounding with water.
Before the mountains had been shaped,
> before the hills, I was brought forth—

when he had not yet made earth and fields,
  or the world's first bits of soil.
When he established the heavens, I was there,
  when he drew a circle on the face of the deep,
when he made firm the skies above,
  when he established the fountains of the deep,
when he assigned to the sea its limit,
  so that the waters might not transgress his command,
when he marked out the foundations of the earth,
  then I was beside him, like a master worker;
and I was daily his delight,
  rejoicing before him always,
rejoicing in his inhabited world
  and delighting in the human race. (Proverbs 8:24-31 NRSV)

John 1. Because of a repetition of words that works in the original Greek but not so well in English, I tried my hand at this translation. The key is to remember that the logos or word "came to be flesh and tabernacles among us" just a few verses later (John 1:14, author's translation).

In the beginning was the Word (logos), and the Word was close with God, and the Word was God. He was in the beginning with God. All things came to be through him, and not one thing came to be without him. What has come to be in him was life, and the life was the light of all people. The light shines in the darkness, and the darkness has not overcome it. (John 1:1-5, author's translation)

Romans 4:17. This is one of the clearest passages describing God's creating out of nothing.

He [Abraham] is our father in the sight of God, in whom he believed—the God who gives life to the dead and calls into being things that were not. (Romans 4:17)

Colossians 1:16 and Hebrews 1:2. I put these two texts together because they are entirely complementary in the way they demonstrate Christ's role in creation.

> In [the Son] all things were created: things in heaven and on earth, visible and invisible, whether thrones or powers or rulers or authorities; all things have been created through him and for him. (Colossians 1:16)

> But in these last days [God] has spoken to us by his Son, whom he appointed heir of all things, and through whom also he made the universe. (Hebrews 1:2)

Revelation 4:11. Creation is an act of God's decision to pour in love and sovereignty. This passage says that better than any other.

> You are worthy, our Lord and God,
>   to receive glory and honor and power,
> for you created all things,
>   and by your will they were created
>   and have their being. (Revelation 4:11)

What We Affirm and Believe

What did you find in these verses? In my mind, these texts can be summarized to affirm something extremely simple: God did it.

Put with just a bit more theological nuance, the doctrine of creation is about the goodness of the God of life—the God who brings things that did not exist into existence with one simple word. It is about the God who continues to create a "new thing" throughout cosmic and human history. It even points to the God who will fulfill creation at the end of time.[11] Or to quote the Apostle's Creed, "I believe in one God, the Father Almighty, Maker of heaven and earth, and of all things visible and invisible." That's the center of our faith—God has made all that is and has made us in his image. And, the biblical texts clearly add, God has made it all good. All was put into its proper functioning at the moment God spoke it into existence.[12]

> "I believe that science and religion can coincide. It just didn't come from nowhere."
> **MICHELLE, AGE 19**

Notice that in all of this, science is affirmed as having a significant contribution—for example, what is the nature of this earth?—but at the same time it is also circumscribed—scientific insight cannot tell us that God made the world, though it may offer us evidence of God's handiwork.

Creation ≠ 6 x 24

A guy starts talking to God and asks, "Hey, God, what does 100 million years seem like to you?"

God answers, "One hundred million years? That's like a second to me."

Then the man asks, "Hey, God, what's 100 million dollars seem like to you?"

"One hundred million dollars? It seems like a penny to me."

So the guy says, "Hey, God, could I borrow a penny?"

And God answers, "Sure. Wait a second."

God's time is not our time. So we need to be careful when we discuss the length of time it took for God to create the world. If we take the consensus of modern science seriously, it's impossible to conclude that God created the universe in six twenty-four-hour days. Still, some persist in doing so. One technique is to impugn modern science. I found this comment on Facebook (the fount of all important ideas) about Christians who accept evolution: "If that has happened, he/she hasn't ever taken every thought captive to Christ and His Word. The Shorter Catechism summarizes what Scripture so plainly teaches: 'The work of Creation is God's making all things of nothing, by the Word of His power, in the space of six days, and all very good.'"[13] In fact, young-earth creationism (YEC) argues that mainstream science is the problem and that embedded in it is atheism. Its adherents' primary motivation is to maintain what amounts to a fairly strict, literalistic approach to the Scripture, which rejects modern science. Since the position I'm taking is that we have to engage mainstream science, I can't find a way to do that and affirm a six-thousand- to ten-thousand-year-old earth. Yet although I'm convinced that YEC isn't true, I don't deny that we are together in the body of Christ.

So how does this position square with the first book of Bible, which describes six days of creation? Some of my college students declare that it doesn't—because Ham and others proclaim that it doesn't—and they leave Christianity behind. Any number of thoughtful people find this position unbelievable. So they simply don't believe it.

In fact, various sciences converge to describe the universe as 13.7 billion years old and the earth as about 4.6 billion years old:

> Geologists have found annual layers in ice that are easily counted to multiple tens of thousands of years, and when combined with radio isotope dating, we find hundreds of thousands of years of ice layers. Using the known rate of change in radio-active elements (radiometric dating), some Earth rocks have been shown to be billions of years old, while the oldest solar system rocks are dated at 4.6 billion years. Astronomers use the distance to galaxies and the speed of light to calculate that the light has been traveling for billions of years. The expansion of the universe gives an age for the universe as a whole: 13.7 billion years old.[14]

Many believers who know science have a fairly straightforward but deceptively simple and satisfactory answer: the six "days" of creation don't equal six days of twenty-four hours each. In fact, the word used primarily in Genesis 1–3 for

"I had a friend who once believed in a ten-thousand-year-old earth, but he lost faith because of science."

SCOTT, AGE 21

"day" is the Hebrew *yom*, which often means "twenty-four hours" but also "a period of time." (Since Hebrew contains a limited number of words, those words need to cover a wide range of meanings.) If indeed you find yourself affirming the science behind the age of the earth and the universe, subscribe to the theory of evolution, and believe that God created and creates this universe, then you generally find yourself under the umbrella of theistic evolution or evolutionary creation. (And then there's also Intelligent Design.)

In many ways we've just considered how to read the book of nature. Now let me offer a few interim remarks on how we can best read the book of Scripture.

Interim Comments on the Hermeneutics

We need a better hermeneutic. I'm willing to claim that if we as the church practiced the following five key principles of biblical hermeneutics, we'd be in a better place to engage with the sciences and to respond to the concerns of eighteen- to thirty-year-olds. They'd make our lives better.

First of all, let's remember that whatever word we use to describe our commitment to the truth and power of Scripture—whether *inerrant*, *infallible*, or anything else—we're committed to the ultimate authority (or primacy) of Scripture and not to our interpretation of it. We do, of course, seek to fulfill 2 Timothy 2:15: "Do your best to present yourself to God as one approved, a worker who does not need to be ashamed and who correctly handles the word of truth." That is the deepest intent of this chapter. However, we are not infallible. So what we say about the Bible we must say with boldness, yes—but also with humility.

Second, along with Old Testament scholar John Walton, we need to affirm that although the Bible can speak to us, it wasn't written for us.[15] We are overhearing a conversation that originally took place with another audience. That means we need to take time to learn about the ancient context in which God spoke in Scripture and then carefully seek to apply that to the twenty-first century. I remember once hearing a sermon by the well-known British preacher and biblical commentator John Stott, who proclaimed that when we consider the footwashing story in John 13, it's about serving and caring and being humble. In the first century AD, that job was washing dirty, stinky feet that walked the roads in sandals. Christ, by example, shows us that the Christian life is about moving to the lowest place, about "downward mobility" (to quote Henri Nouwen) and humbling ourselves as Paul demonstrates in Philippians 2:1-11. Similarly, the cosmology of the Bible doesn't sound like 14 billion years because—if we can understand God's mind—that extended timeframe wouldn't have made sense to ancient Hebrews or first-century Christians. The Bible is written in the thought forms of its day.

Third, as my New Testament professor Joel Green used to say, "a text without a context is a pretext." We can't simply hope to apply a passage directly to our context without taking in the original setting. So learning the historical, social, and scientific context of the passage can keep us from a myriad of interpretive sins. In other words, we have to work hard to find what's in the passage and not what we'd like to find. We need to avoid making Scripture a mirror of our own faces and convictions. One way we can do this is to take in the entirety of a passage when we're reading a verse, and the entirety of book when we're reading a chapter, and the entirety of the biblical witness when we're reading a book.

Fourth, the literal interpretation is not the only one. That is to say, we need to ask ourselves what the nature of the text is. If it's a poem, read it like a poem, not like a doctrinal treatise sent from the papal offices. If it's a parable, it's fine if there's exaggeration. In fact, we should expect it. (Who, for example, walks around with a log in their eye as Jesus describes in the brief parable of Matthew 7:3?) If it's a historical passage, read it as history and remember that in earlier times, history and interpretation were often wound together. In the case of Genesis

1–3 and other similar texts, I'm not the first to conclude that it reads like a poetic or liturgical text, not a scientific one.

One final sagacious piece of hermeneutical advice I treasure to this day came through the sermons I heard as an undergraduate at First Presbyterian in Berkeley. Earl Palmer taught us repeatedly that *lean is better than luxurious*. (This parallels the scientific principle of parsimony.) Go for the leaner, more humble interpretation. And when we don't know, it's okay to say that too.

In case my point has become obscure, let me return to it: Emerging adults who sense that the Bible is outdated in light of science can still read it with confidence. They can be convinced by the truth of both of the two books, but the two aren't the same. For one thing, the book of nature doesn't include major theological tenets, like the incarnation of God in Christ. This means we can't make them of equal importance.

We have been considering the creation of the universe, and the next chapter offers two ways we can read too much from the book of nature as we interpret scientific cosmology.

MAKING TOO MUCH OF A GOOD THING

BIG BANG AND FINE-TUNING

The effort that tries to resolve conflicts between Christianity and science by stating that religion is about persons and not about the universe of things should seem a very poor half truth. For God, at least the Christian God, is above all the Creator of the Universe.

STANLEY L. JAKI

We need to move into two areas of mainstream science that connect with the first article of Christian faith. Specifically, how does our doctrine of God as Creator intersect with Big Bang cosmology and cosmic fine-tuning? Repeatedly in the first chapters of Genesis, God calls his creation "good," and I think it's reasonable to observe that the doctrine of creation is good news. But also, can we make too much of a good thing—specifically when it comes to the Big Bang and cosmic fine-tuning?

First we need to examine the theological concept of creation out of nothing (or *creatio ex nihilo* in Latin)—there was no thing when God created. God didn't order a universe that was already there.

God hasn't coexisted with nature. Instead God is entirely sovereign over the heavens and the earth. "For [God] spoke, and it came to be" (Psalm 33:9). The play didn't begin with a stage already there, with the actors walking onto the stage. The author of the play made everything—the actors, stage, the seats, the lights, the theatre, and so on, including the very existence of all material reality.

This brings us to Big Bang cosmology, an extrapolation of Albert Einstein's 1916 theory of general relativity, which pointed to a universe that isn't static (in contrast to what Einstein would have liked). Instead relativity implies that the universe is expanding. When expansion is extrapolated in reverse, this implies an initial singularity when time began, or t = 0. This is the "'day without yesterday,' where space was infinitely curved and all energy and all matter was concentrated into a single quantum of energy."[1] The expanding cone of the universe has a point at one end, and that point is where we can locate the Big Bang.

Because Einstein preferred a static universe, George Lemaître, the Belgian Jesuit priest and mathematician, and Edwin Hubble, the maestro of astronomical investigation, had to convince Einstein he was right. Einstein, however, was not the only scientist in need of persuasion. The brilliant Cambridge astronomer Fred Hoyle presented the reigning view of the time, "steady state" cosmology, with two colleagues, Herman Bondi and Thomas Gold, in 1948. Simply put, they said the universe was virtually unchanging and of infinite age. The cosmos always was, and always would be, the same. Interestingly, this sounds like Spinoza's metaphysics, but Hoyle looked to another philosopher because he did not like the implications of creation out of nothing for theological and philosophical reasons.

> Unlike the modern school of cosmologists, who in conformity with Judaeo-Christian theologians believe the whole universe to have been created out of nothing, my beliefs accord with those of Democritus who remarked, "Nothing is created out of nothing."[2]

Hoyle argued forcefully for his theory and made sure to disparage the alternative with a slight. He named it the "Big Bang."

In the 1950s and '60s, these two theories contended for scientific approval. It was not until 1965 that Robert Wilson and Arno Penzias of the Bell Laboratories detected background static radiation in the cosmos. This static was so unexpected that Wilson and Penzias believed pigeons had roosted in their enormous instruments. In fact, the pigeons had made a home there and even left some droppings. Once all that was purged, however, the background static remained. It was the "echoes" of that initial explosive moment of the Big Bang. The COBE satellite probed outer space in 1989 and found further confirmation of this background radiation through its far infrared absolute spectrophotometer. All told, Big Bang cosmology has won out.

These discoveries provide an amazing consonance with Paul's words in Romans 4:17 about the God who "calls into being things that were not"—*creatio ex nihilo*. The late Robert Jastrow certainly thought so. This astronomer, planetary physicist, and self-described agnostic offered these somewhat hyperbolic comments on the Big Bang cosmologist, with a view to Genesis 1–3:

> A sound explanation may exist for the explosive birth of our Universe; but if it does, science cannot find out what the explanation is. The scientist's pursuit of the past ends in the moment of creation. . . . This is an exceedingly strange development, unexpected by all but the theologians. They have always accepted the word of the Bible: In the beginning God created heaven and earth. . . . For the scientist who has lived by his faith in the power of reason, the story ends like a bad dream. He has scaled the mountains of ignorance; he is about to conquer the highest peak; as he pulls himself over the final rock, he is greeted by a band of theologians who have been sitting there for centuries.[3]

I quote this because, honestly, it's fun for me as a theologian. It is, however, overstating the case.

Yes, theologians have long held that creation happened "at once" from nothing, and they can be flattered by Jastrow's conclusions. It's not clear, however, whether to have a "beginning" means to be "created,"[4] whether t = 0 is really the initial act of creation, or whether, perhaps, it's one in a cycle of expansion and contractions of which there have been, and will be, others.

There are also competing forms of scientific cosmology that upend Big Bang. Scientist and atheist Lawrence Krauss recently presented one alterative to divine creation based on the effects of quantum "nothingness."[5] But quantum nothingness is not really the absence of existence. As I wrote above, the doctrine of *creatio ex nihilo* is that there was no thing, not even quantum nothingness, "before" creation. (I set "before" in quotation marks because God's creation also started time. So there really isn't a temporal before and after.) Instead of Krauss, let us take note of James Hartle and Stephen Hawking's model describing the coming together of "three spaces" to create the four-dimensional spacetime universe as a "bowl" instead of a "cone," and thus there is no t = 0.[6] All this is quite intricate, so I'll leave the details aside except to say that there are competing paradigms for cosmology, and certainly more are in the works.

> "Science and nature were fascinating for me. I wondered why things happened—how we have what we have. I loved watching things grow. It opened up all kinds of questions. How did this get created?"
>
> *CHRISTINA, AGE 20*

As Christians, we can affirm that science, for now, and Christian faith find consonance at t = 0. Still, I advise caution when it comes to putting our faith in the Big Bang—that is, from reading directly off the page of the book of nature as if it's the book of Scripture. We don't put our faith in scientific discoveries, but in the God who created—whether by Big Bang or not.

This Big Bang appears to be an extremely precise event. And this leads us to cosmic fine-tuning, or the anthropic principle. Oft-maligned Wikipedia actually offers an excellent definition of cosmic fine-tuning:

> The conditions that allow life in the Universe can only occur when certain universal fundamental physical constants lie within a very narrow range, so that if any of several fundamental constants were only slightly different, the Universe would be unlikely to be conducive to the establishment and development of matter, astronomical structures, elemental diversity, or life as it is understood.[7]

Given this definition, it's easy to see why this concept is called "fine-tuning," but why is it also referred to as the "anthropic principle"? Since the 1960s, when scientists began to identify discrete, precisely calibrated parameters that produced the universe we know, they also hypothesized that these conditions particularly allowed for the emergence of conscious, moral creatures. "Anthropic" comes from the Greek word for human being, *anthropos*; the anthropic principle states that the universe is fitted from its inception for the emergence of life in general and intelligent, moral life in particular, though not necessarily earthly human beings like you and me.[8]

Cosmic fine-tuning seems like a decisive victory for God's creation and against the common atheist assertion that we are the products of blind, random, impersonal forces. These perfectly calibrated, fine-tuned parameters have led many believers to nod in agreement with Freeman Dyson, the physicist who spent many years at Princeton's famed Institute for Advanced Study, when he says, "The more I examine the universe and the details of its architecture, the more evidence I find that the universe in some sense must have known we were coming."[9] This seems conclusive to many Christians. Fine-tuning offers evidence in the structure of creation for the fingerprint of God.

But is it a proof for God? Here's where it's easy to overstate the case. Some present the multiverse theory as a rejoinder—in other words, there have been innumerable attempts at other universes that simply failed. But since most of my colleagues in science tell me this theory is metaphysical speculation because it is in principle inaccessible to our scientific verification, I turn to philosophy for the strongest argument against fine-tuning as a proof for God's creation: it's a tautology. Simply put, we are here in this particular universe. Whether its existence is perfectly calibrated or not, it's the only universe we've got. It's the only universe we know. In a sense, that makes the probability of its existence one hundred percent. Similarly, it's just as intrinsically improbable that a person named Greg is typing on a MacBook Pro at his home in Chico on the fine-tuning argument because he's thinking about the doctrine of creation and science. But yet here I am. And we don't offer that set of data as evidence for a designer.

This is not a deductive proof for God that leaves no room for disagreement. Instead it's a suppositional argument that offers confirmation for the judgment that this universe has design and that design is confirmed, to some degree, by the incredible particularity of its parameters. If we *suppose* there to be a God who desired the universe, we should expect that this universe would have evidences of the design. The fine-tuning of various physical constants is therefore consistent with God's design. Therefore it is reasonable to assert that God exists.

Here's an analogy. Suppose that tonight is my wedding anniversary. First scenario: When Laura arrives home, I declare, "Laura, I've been planning to celebrate this anniversary big-time!" I immediately call a pizza company to deliver, grab a random assortment of napkins, glasses, and plates (most of the dishes are dirty), and fumble through my iPod for background music. Second scenario: When Laura leaves work, a limo picks her up, and I am in the back seat pouring Veuve Clicquot into luxurious champagne flutes. I toast, "Here's to our anniversary!"

We arrive home, where a chef is set to serve dinner at our house on a candlelit table set with china and crystal while a string quartet plays in the background. Sounds nice, right?

Which of the two scenarios has more specific parameters and therefore better supports my contention that I really intended to celebrate my anniversary? It's obviously the second.

Alister McGrath, Oxford professor of theology and science, can offer the last word. The cosmological factors highlighted by cosmic fine-tuning don't offer "irrefutable evidence for the existence or character of a creator God." Instead, he writes:

> What would be affirmed . . . is that they are consistent with a theistic worldview; that they reinforce the plausibility with the greatest ease within such a worldview; that they reinforce the plausibility of such a worldview for those who are already committed to them, and that they offer apologetic possibilities for those who do not yet hold a theistic position.[10]

In conclusion, here's a summary of the principles discussed above.

- Seeking integration of science and faith doesn't mean seeing the book of nature and the book of Scripture as identical.
- Be careful of making too much of science because the results of science change.
- Keep up with the developments of science as best you can to avoid bolting your theological conclusions onto an older scientific theory.
- It's probably good to let any major scientific theory bake for five to ten years before it's ready for theological consumption.

All of this brings us the questions we encounter in Genesis and Romans on Adam and Eve. Let me apply those hermeneutical principles from chapter four to the question of whether Adam and Eve were historical.

5

ADAM, EVE, AND HISTORY

*I can show you your lineage back to Adam—
that's all you need to know.*

**PARENTS OF A COLLEGE STUDENT WITH
QUESTIONS ABOUT EVOLUTIONARY BIOLOGY**

Sometimes called "the world's shortest poem," this little verse by Ogden Nash sets out in almost ridiculous simplicity some provocative questions:

Fleas
Adam
Had 'em

Just in case the poem's depth of insight is opaque, here are the questions it raises for me: Did this first human and his wife, Eve, experience anything annoying as fleas? Was that before the fall in Genesis 3 or after? Was there really a historical time "before" and "after" the fall where Adam and Eve ate the fruit and everything changed, including irritating realities like fleas?

I think you see the point. What do the Scripture and science tell us about Adam and Eve and whether they are historical? How do we bring scientific discoveries to bear on the biblical witness? My conviction is that Christians need to embrace, not fear, mainstream science, and that this work is critical for emerging adults.

Mainstream science, however, has work to do in convincing many evangelicals about some of its conclusions. According to a 2012 Pew Report, many Christians do not believe that human beings evolved. "A majority of white evangelical Protestants (64%) and half of black Protestants (50%) say that humans have existed in their present form since the beginning of time." Interestingly, the report adds that 78 percent of white mainline Protestants are in support of human evolution.[1] One factor, for those of us who take the Bible seriously as God's Word to us, is that human evolution is hard to square with a literal Adam and Eve. (By the way, too many discussions leave out the Eve part here, but I think she's important. So I'll work to keep her in the picture.)

> "I don't think we came from monkeys, but Adam and Eve is weird too."
>
> *LEXI, AGE 19*

Let's look now at key sections from Genesis 2–3.

Then the LORD God formed a man from the dust of the ground and breathed into his nostrils the breath of life, and the man became a living being. (Genesis 2:7)

The LORD God took the man and put him in the Garden of Eden to work it and take care of it. And the LORD God commanded the man, "You are free to eat from any tree in the garden; but you must not eat from the tree of the knowledge of good and evil, for when you eat from it you will certainly die."

The LORD God said, "It is not good for the man to be alone. I will make a helper suitable for him."

Now the LORD God had formed out of the ground all the wild animals and all the birds in the sky. He brought them to the man to see what he would name them; and whatever the man called each living creature, that was its name. So the man gave names to all the livestock, the birds in the sky and all the wild animals.

But for Adam no suitable helper was found. So the LORD God caused the man to fall into a deep sleep; and while he was sleeping, he took one of the man's ribs and then closed up the place with flesh. Then the LORD God made a woman from the rib he had taken out of the man, and he brought her to the man. (Genesis 2:15-22)

Now the serpent was more crafty than any of the wild animals the LORD God had made. He said to the woman, "Did God really say, 'You must not eat from any tree in the garden'?"

The woman said to the serpent, "We may eat fruit from the trees in the garden, but God did say, 'You must not eat fruit from the tree that is in the middle of the garden, and you must not touch it, or you will die.'"

"You will not certainly die," the serpent said to the woman. "For God knows that when you eat from it your eyes will be opened, and you will be like God, knowing good and evil."

When the woman saw that the fruit of the tree was good for food and pleasing to the eye, and also desirable for gaining wisdom, she took some and ate it. She also gave some to her husband, who was with her, and he ate it. Then the eyes of both of them were opened, and they realized they were naked; so they sewed fig leaves together and made coverings for themselves.

Then the man and his wife heard the sound of the LORD God as he was walking in the garden in the cool of the day, and they hid from the LORD God among the trees of the garden. But the LORD God called to the man, "Where are you?" (Genesis 3:1-9)

Genesis 3:9 may be the most chilling verse of the Bible. God is so disturbed by human disobedience that the one known as omniscient asks, "Where are you?" That summarizes the fall—the amazing, terrifying distance between us as creatures and our Creator. The other curses (like sweat of the brow and increased pain in childbirth) are terrible yet derivative. This frightening separation can be outdone only by Jesus' words on the cross when he takes on human sin: "My God, my God, why have you forsaken me?" (Matthew 27:46).

In this story's simplest framing, all Christians believe that Adam is a type of Christ—thus Adam is typological—but do he and Eve also have to be historical? Let me break the relevant issues surrounding the questions and answers about the historical Adam and Eve into three sentences:[2]

- *God specially and suddenly created a literal first pair of humans (Adam and Eve).* The third- or fourth-century BC book of Tobit describes this clearly: "You created Adam and gave him his wife Eve to be his helper and support. They became the parents of the whole human race" (8:6).

- *All people are biologically descended from these two persons.* This is a contentious statement in contemporary thinking. We will spend more time on it in this chapter.

- *Because of the actions of these two humans, all human beings are under the power of sin and in need of salvation.* This is, broadly speaking, the doctrine of original sin as well as the universal need for salvation. The choice Adam and Eve made in the Garden of Eden affects all human beings throughout time—and not in a good way.

In response to these questions surrounding Adam and Eve, there are also three dominant positions by mere Christians. Imagine these three on a continuum with position one taking the most literal interpretation of the relevant biblical texts, position three opting for a natural, often typological, but not literal interpretation, and position two mediating between them.

On Adam and Eve

Position one is embraced primarily by young-earth creationism (YEC). It maintains a traditional perspective of a relatively young earth (about six thousand to ten thousand years old) and God's special creation of a historical duo, Adam and Eve, who lived in a perfect state for a period of time, ate a fruit (not particularly an apple—that detail belongs to seventeenth-century poet John Milton), and experienced condemnation for their sin and death. This position rejects mainstream science. For this reason I'm convinced it's not true, and I won't have much more to say about it. (I don't, of course, deny that those Christians who hold to YEC are members of the body of Christ. I simply am persuaded they are wrong about their science and thus components of their theology.)

Other "mere Christians," such as C. S. Lewis, propose position three: a literal Adam and Eve never existed; instead that they are paradigmatic of the human condition. Lewis wrote, "For long centuries, God perfected the animal form which was to become the vehicle of humanity and the image of himself."[3] In this view we are not descended from one pair of humans but from the gradual evolutionary development of hominins.[4] Thus we share common descent with the great apes—but please note this does not mean we "descended from monkeys." According to evolutionary theory, great apes and humans arose from a single ancestral species that existed in the distant past. From that common ancestor, separated populations developed in various directions and ultimately split off to form the different hominid species that are alive today. Lewis contended that God implanted a divine consciousness in those early animals or hominins, but

> we do not know how many of these creatures God made, nor how long they continued in the Paradisal state. But sooner or later they fell. . . . They wanted, as we say, to "call their souls their own." But that means to live a lie, for our souls are not, in fact, our own.[5]

This means we all are created for good and we all turn away, but there was no one historical first pair specially created by God out the dust and then from a rib (not through the normal process of childbirth).

Does this position have biblical support or is it simply a capitulation to science? In Genesis 1–3, the word *adam* in Hebrew simply means "human" generically and is often used in these texts with the article "the"—thus "the human." *Eve* means "mother of the living" or "life." Thus their proper names were not "Adam" and "Eve" but essentially "Human" and "Life"—or, as New Testament scholar Scot McKnight phrases it (emphasizing nuances in the original Hebrew), "Dusty" and "Mama."[6] Moreover, if this pair lived even six thousand years ago, it is improbable that they spoke Hebrew, which didn't come into existence until at least a couple thousand years later. All in all, Adam and Eve don't really come off as proper names but as symbolic or typological ones. This is a signal that, based on the nature of the text, we shouldn't interpret them literally. (This means I'm taking a *natural* over a *literal* interpretation.)

Position three lines up easily with modern science. For example, it squares with modern genetic studies of populations, which do not support the belief that we are all descended from one sole pair.[7] But it does not endanger the core elements of the gospel (more on that below). I call this position typological, paradigmatic, or archetypal. *Typological* indicates that the first Adam is a type of human being (see Romans 5:14) and thus typical of human experience. Similarly, *paradigmatic* means that what he and Eve experienced—of being called to, and moving away from, God—is a pattern or paradigm for every human. Many use the word *archetypal* because Adam "embodies all others in the group," as John Walton puts it.[8] In any event, sin enters when we move toward self and away from God. We pay the price not for a human pair's transgression back in history but for our own individual and collective sin. Thus we need the Redeemer, the historical figure who died on a cross under a particular Roman procurator, Pontius Pilate, and who rose bodily from the dead. The point of all this, as New Testament scholar N. T. Wright says, is "the call to be an image-bearing human being renewed in Christ."[9]

To this, some will respond, "Yes, position three lines up easily with modern science—too easily." And so we come to position two, which falls somewhere between YEC and a typological but nonhistorical Adam and Eve. In summary, this view takes in modern scientific consensus on the age of the earth and development of hominins but says, "Hey, wait! We can't simply jettison Adam and Eve as real, historical people. There are biblical and theological commitments that are wrapped up in this." Position two is convinced that Adam and Eve are in some ways historical figures (this is the view of John Walton, S. Joshua Swamidass, C. John Collins, and Tim Keller), but generally sets out a period of time for common descent with other primates and then designates a point when God decided to set Adam and Eve apart as the first and original image-bearing *Homo sapiens*. The period of time between Adam and Eve and us varies. Swamidass, professor of laboratory and genomic medicine at Washington University in St. Louis, has called attention to genealogical science, which demonstrates the plausibility that we all share common genealogical ancestors very recently. (He also concludes that there is no evidence against the special creation of Adam, ancestor of us all, within a larger population.)[10]

The evangelical biblical scholar Derek Kidner, in his commentary on Genesis, proposes that the two kinds of humans at the time of Adam could be called "Adamites" and "pre-Adamites."[11] Kidner's "tentative" concept could fit with geneticists' theory of human origination from a single larger population. He proposes that pre-Adamites and Adamites shared the same genetic heritage and existed simultaneously. There was, however, "no natural bridge from animal to man." God had to place his image upon Adam, and then he may have acted similarly with the others who existed at that time "to bring them into the same realm of being." In Kidner's view, Scripture presents Adam's sin "in terms not of heredity but of solidarity." In theological terms, his "federal headship" extended "outwards to his contemporaries as well as onwards to his offspring, and his disobedience disinherited both alike."[12]

Someone who summarizes these positions well and knows how to take the biology seriously is my friend and colleague Gary Fugle, a career professor of biology. Gary is fully Californian—you can imagine him grabbing a volleyball at the beach and saying "dude" with no irony. He grew up in a "lightly Christian" Presbyterian household; his mother occasionally went to church while his father stayed home. In high school he was "more interested in sports and girls than science," he told me. While at the University of California, Santa Barbara, he fell in love with the study of biology—he even won the top biology student award—and eventually went on to earn a PhD in evolutionary biology. At UCSB he met some thoughtful Christians. They didn't necessarily answer his every question, but they had a confidence that what was true (whether discovered in the Bible, through the sciences, or in other ways) was good and pointed to Christ. Gary also found the "wisdom of St. Clive" (my words) when he read *Mere Christianity* and began to see the reasons for Christian faith.

> "Hearing Gary Fugle's lecture . . . was absolutely eye-opening for me. His lecture opened my mind completely to the idea that it is possible to be a man of faith as well as science."
>
> **SUSAN, AGE 19**

Then a marvelous serendipity occurred: in studying evolutionary biology, Gary became a Christian. He taught biology his entire career (before retiring a few years ago) while maintaining leadership roles in his evangelical Presbyterian church.

There are different paths to Christian faith. I was particularly struck by the awe, wonder, and beauty I experienced sitting in the solitude of a high mountain lake, gazing at a colorful sunset, or pondering the delicate structure of a flower. Atheistic scientists argue that these universal emotions are solely products of a mechanistic evolutionary history. I am convinced otherwise: in these responses we are sensing God's hand in his creation.[13]

You can't say—or at least I can't—that Gary doesn't take the Bible seriously. He does. But he is also convinced by evolutionary science. He sees the development of *Homo sapiens* (our ultimately image-bearing species) occurring over time. And he presents two possibilities for the historical Adam that square with modern science. One, "that Adam was singly taken aside by God from physically evolved humans and the image was divinely imparted to him." He adds that this image "was not something that simply evolved along with human physical features."[14] The second possibility is that God "revealed himself in a special way to two individuals or a group of humans and this knowledge of God spread outward to other people who would hear."[15] This latter idea solves some puzzles—for example, where the wives came from for Cain and Seth without violation of God's prohibitions against incest and the population of other human beings implied in Genesis 4.

I have to admit that this position doesn't really sound to me like a historical Adam and Eve—like two people that God created specially in a garden in Mesopotamia six thousand or so years ago. This is why some hold to a typological Adam and Eve. Position two takes Adam and Eve seriously but leaves aside other considerations. For example, Genesis 2:7 says, "the LORD God formed a man [adam] from the dust of the ground and breathed into his nostrils the breath of life." It doesn't take a strict literalist to conclude that this isn't describing the birth of Adam from a woman. The text describes Adam and Eve's creation directly by God. He doesn't have a father and mother, but in position two he would. In other words, as a pastor commented in one of the monthly science and religion meetings I host, "Why interpret some components of Genesis 1–3 literally but not others?"

"I wasn't taught as a child about evolution. I first learned about it in ninth grade. . . . It's hard to believe what I can't see."

MICHAELLA, AGE 19

There are several other biblical texts to consider, but one is absolutely critical—and it leads into an interconnected doctrine, namely, original sin. How is it that Adam and Eve's disobedience affects us today? Even more, as Wright would want me to include, how is it that the divine calling of image-bearing was lost and ultimately reestablished in the person and work of Jesus Christ? This concept is so critical that we need to look at Romans 5:12-21 in full:

> Therefore, just as sin entered the world through one man, and death through sin, and in this way death came to all people, because all sinned—[16]
>
> To be sure, sin was in the world before the law was given, but sin is not charged against anyone's account where there is no law. Nevertheless, death reigned from the time of Adam to the time of Moses, even over those who did not sin by breaking a command, as did Adam, who is a type of the one to come.
>
> But the gift is not like the trespass. For if the many died by the trespass of the one man, how much more did God's grace and the gift that came by the grace of the one man, Jesus Christ, overflow to the many! Nor can the gift of God be compared with the result of one man's sin: The judgment followed one sin and brought condemnation, but the gift followed many trespasses and brought justification. For if, by the trespass of the one man, death reigned through that one man, how much more will those who receive God's abundant provision of grace and of the gift of righteousness reign in life through the one man, Jesus Christ!
>
> Consequently, just as one trespass resulted in condemnation for all people, so also one righteous act resulted in justification and life for all people. For just as through the disobedience of the one man the many were made sinners, so also through the obedience of the one man the many will be made righteous.

The law was brought in so that the trespass might increase. But where sin increased, grace increased all the more, so that, as sin reigned in death, so also grace might reign through righteousness to bring eternal life through Jesus Christ our Lord.

Secondarily we have Paul's succinct formulation in 1 Corinthians 15:21-22:

For since death came through a man, the resurrection of the dead comes also through a man. For as in Adam all die, so in Christ all will be made alive.

One man, the first Adam, paired with one man, the new Adam. This seems fairly clear—Paul believed in a historical Adam and so should we. Here's why many evangelicals will find themselves uncomfortable with the purely paradigmatic or typological approach. It doesn't seem to square with the natural reading of Scripture and certainly contradicts many statements of evangelical faith, as summarized in Wheaton's statement: "God directly created Adam and Eve, the historical parents of the entire human race."[17]

This interpretation also fits with the classic view of original sin, which Augustine put his stamp on and which flows from his translation of the final phrase from verse 12. Instead of "because all sinned," Augustine used the Latin translation of the original Greek (called the Vulgate), "in whom all sinned." Adam did it, and in him we also all sinned. The original Greek phrase "because all sinned" seems to leaves a less direct connection and affirms that we are all guilty because we all commit sin. Moreover, many commentators seem to miss the apocalyptic language of Paul's letters. Here's what I mean: Adam and Eve's fall initiated a cosmic change. It defines an old age or world that is passing away (1 Corinthians 7:31). With Christ a new world (or age) has begun. Christians from Paul's time until today are caught between these two worlds, between their power. We feel torn as Paul describes in

Romans 7:13-25,[18] and so we groan with all creation for the full revelation of what the new Adam, Christ, has brought (Romans 8:18-24);
indeed the Spirit intercedes for us with groanings too deep for words
(Romans 8:26-27).

With all of that in mind, which position best fits biblical studies, theology, philosophy, and science? The noted evangelical scholar and biblical
commentator James D. G. Dunn interprets Romans 5, especially verse 12,
with a nuance that fits with a fully typological or paradigmatic approach:
"The reference to Adam's failure is for Paul a way of characterizing the
condition of humankind in the epoch of human history which has
extended from the beginning of the human race till now."[19] He continues:

> In particular, it would not be true to say that Paul's theological
> point here depends on Adam being a "historical" individual or on
> his disobedience being a historical event as such. Such an impli-
> cation does not necessarily follow from the fact that a parallel is
> drawn with Christ's single act; an act in mythic history can be
> paralleled to an act in living history without the point of
> comparison being lost. So long as the story of Adam as the
> initiator of the sad tale of human failure was well known, which
> we may assume (the brevity of Paul's presentation presupposes
> such a knowledge), such a comparison was meaningful.[20]

Dunn summarizes by emphasizing the centrality of what Jesus Christ
has done: "Indeed, if anything, we should say that the effect of the
comparison between two epochal figures, Adam and Christ, is not so
much to historicize the individual Adam as to bring out the more than
individual significance of the historic Christ."[21]

Dunn's words carry us to a key point we ought not miss: Christ is the
focus of Paul's letters—and indeed the whole New Testament—not
Adam. Therefore if Adam proves to be nonhistorical but solely paradigmatic, there's no problem for what Paul is teaching here. Position three

concludes, "Yes, Paul didn't teach the historical Adam as doctrine. So this is good exegesis, faithful to an orthodox confession of Christ, and matches with excellent modern science."

Position two cannot agree. I can imagine that few would want to differ too readily with former New York City pastor and bestselling author Tim Keller (I know I don't):

> [Paul] most definitely wanted to teach us that Adam and Eve were real historical figures. When you refuse to take a biblical author literally when he clearly wants you to do so, you have moved away from the traditional understanding of the biblical authority. . . . If Adam doesn't exist, Paul's whole argument—that both sin and grace work "covenantally"—falls apart. You can't say that "Paul was a man of his time" but we can accept his basic teaching about Adam. If you don't believe what he believes about Adam, you are denying the core of Paul's teaching.[22]

So how do we decide where to place ourselves on this spectrum of views of the historical Adam?

Let's Keep Our Eyes on Jesus

To answer that last question, let me repeat: the center of our faith is Christ, not Adam. It's worth noting that Adam does not make extensive appearances in the Bible nor the creeds.[23] Which views fit with the work of the historical God-man Jesus the Christ and his offer of redemption through his life, death, and bodily resurrection? Let's keep our eyes on Jesus as a real historical figure. He is our center. Therefore, we have to start with Jesus Christ—with his life, death, and resurrection, with the fact that he has saved us from sin, the world, and the devil— and then see what this implies about Adam.

Again to Wright: when we hear the name Adam, let's not only think "original sin" but remember that Adam's vocation was to rule over all creation.[24] He is the paradigmatic "man" or "humankind" who is celebrated in Psalm 8:

> When I consider your heavens,
>   the work of your fingers,
> the moon and the stars,
>   which you have set in place,
> what is mankind that you are mindful of them,
>   human beings that you care for them?
>
> You have made them a little lower than the angels
>   and crowned them with glory and honor.
> You made them rulers over the works of your hands;
>   you put everything under their feet. (Psalm 8:3-6)

This in fact returns me to my big affirmation: as Christians, we believe that redemption comes through the grace of Jesus Christ, received by faith. This is the universal answer for the universality of sin. It is also the best way to approach original sin. In his magisterial and thorough review of the sources for the doctrine of original sin, Henri Rondet reminds us that even there the emphasis is on what Christ has done: "For a Christian of the very first few centuries original sin was not in the foreground; on the contrary, the redemption was the fundamental assertion."[25] Or again, as Tatha Wiley writes in her historical and systematic summary of the doctrine of original sin: "The concerns of the early Christian writers were soteriological—they emphasized what God had done for humankind through Christ. At the heart of their proclamation was the conviction that Christ had overcome the estrangement from God caused by sin."[26]

This reminds me of physicist Karl Giberson's quip that our belief (or not) in a historical Adam and Eve "shouldn't cause us to hurl accusations of infidelity at one another."[27] So if some of these views about

Adam make a portion of us in the church nervous—believing that if the other side wins we'll soon jettison all biblical truth or all engagement with modern science—let's remember that the process goes both ways. Here let's start not with theologians or biblical scholars but with the science philosopher Imre Lakatos, who maintained that certain teachings at the "hard core" of Christianity (the divinity of Christ as the historical God-man, for example) are not jettisoned easily, even in the presence of anomalies.[28] Position three holds that Adam and Eve's historicity is not part of the "hard core" of Christian faith (in Lakatos's words), while position two (and one) hold that it is.

The bestselling author and theologian Greg Boyd read Lewis's views as an undergraduate while he was struggling with young-earth creationism, which he believed to be the Christian consensus. He thought YEC made little sense scientifically, and Lewis's insights into Adam and the fall helped him keep the Christian faith viable. Ultimately, Boyd was inclined to believe in a historical Adam, but the experience of reading Lewis and the purely typological view led him to this conclusion: "I, as a pastor of an evangelical and Anabaptist church, think it vitally important that we *not* put forth the historicity of Adam as a matter that is essential to Christian faith."[29] Regarding those who can't see harmony between the statements "I believe in Christ" and "I don't believe Adam was historical," he says, "I implore them to refrain from becoming dogmatic on this point and simply to trust the genuineness of those who disagree. The fact is, dogmatism on this point would have tragically barred C. S. Lewis, myself, and a multitude of others from the life-giving kingdom."[30] This debate, he concludes, "should be construed as a debate among orthodox Christians, not as a debate that determines whether or not one is an orthodox Christian."[31]

Let's also remember that emerging adults are fatigued by conflict, especially among Christians. For the demographic that is the focus in this book, positions two and three both take mainstream science and mere Christianity seriously. Though we can certainly disagree and discuss why, let's do so without threatening excommunication. For these reasons, Boyd's words are wisely irenic. From one Greg to another, I say, Amen!

Hermeneutics in a Scientific and Technological Age

All of the discussion above arrives at the church's door because we live in a scientific and technological age, which leads me to return to our belief in the Bible and why it can still speak today. All in all, the church needs a better hermeneutic, a method of biblical interpretation that reads the Scripture faithfully and mindfully. Here are five reflections on what that looks like.

We hold to the Bible because there we find our relationship with God through Jesus Christ. The book of Scripture, not the book of nature, witnesses to Christ. This leads to a gauntlet laid down by scientific progress. Should we scrap Scripture in our scientific age, or at least place a significant boundary around its use? (This, of course, is what many modern critical approaches to the Bible seek to do.) Here we could distinguish between the speculative thought that dominates science and the logic of personal relationships that is essential to belief in, and relationship with, God. Should we give up the Bible in light of science? No, because our primary motivation with Scripture is a personal relationship with God, the author of Scripture. We stay in relationship even if there is evidence to the contrary. So too with the Scripture where we meet Jesus Christ. As Jesus tells the Jewish leaders in John 5:39, "It's the Scriptures that give witness to me" (my translation). In Martin Luther's words, "The Bible is the cradle wherein Christ is laid."[32] Therefore, we are rightfully hesitant to abandon it.

I certainly realize, as the philosopher of science Michael Polanyi emphasizes, that there is also a personal engagement with science—his epic book on science is, after all, named *Personal Knowledge*.[33]

Among other factors, scientists are people who do science. None-theless, the truths of science are often necessary ones—like 2 x 2 = 4. That is not something that's self-involving like the claim, the demand, the offer of the gospel that in order to be forgiven for our sin we need to trust in Jesus Christ. Scripture in general is not a mathematical truth but a self-involving one. It's closer to engaging a literary text than an arithmetic one.

Although we seek integration, we need to interpret Scripture with a sufficient dose of independence between science and faith when necessary. Galileo may not have been entirely right with his quip, "I would say here something that was heard from an ecclesiastic of the most eminent degree: 'That the intention of the Holy Ghost is to teach us how one goes to heaven, not how heaven goes.'"[34] But he was on to something. When Psalms 8 and 19 lead us to consider the heavens and the glory of humankind, they don't tell us how to use a telescope, in-terpret the mathematics of physics, or understand comparative anatomy. These are all human endeavors complementary to the study of Scripture. If good science leads us to an earth that's 4.5 billion years old and human beings who have evolved, then we need to follow it. As theologian Harold Nebelsick commented so poignantly, "To ignore the discussion of today's science is simply to discuss theology in terms that are related to the science of the by-gone era."[35] And that means we believe but lose our minds in the process.

The interests of the interpreter are critical to the task of interpre-tation. This isn't just "we find what we're looking for," but Christians reading the Bible to enhance their prayer lives will come to different results than historians seeking to understand historical events in the life of ancient Israel. In science this is critical rather than naive realism; the world exists whether we perceive it or not, but the way it's understood depends to some degree on the observer. This is something physicists learned from the famous two-slit experiment—depending on the way an experiment is set up and what scientists are measuring, photons can behave either as waves or as particles. This leads to the inverse, and paradoxically, it's a grand positive.

Science is not the sole arbiter of truth. Put another way, the Bible is not a scientific textbook. It doesn't claim to be and doesn't need to be. What I'm hoping to accomplish here is to offer the contours in which we can take the biblical texts as a unique authority by not requiring them to come under some interpretive structure that mimics empirical science. I lean on a quotation that's been ascribed to Albert Einstein but apparently came from William Bruce Cameron: "Not all that counts can be counted."[36] Indeed, we do not need to provide a quantitative approach to Scripture but instead read it as God's word to the church.

Our biblical interpretation is about learning to live within the narrative of the Scripture. This is letting God's story become our story, as it were. We don't memorize the Bible as we do the Periodic Table. It's not mathematical formulas. Again to cite St. Clive: we receive God's Word in the Bible "not by using it as an encyclopedia or encyclical but by steeping ourselves in its tone and temper and so learning its overall message."[37] The message is that God has created it all good, that God has created us to bear the divine image, and that Christ has restored this vocation.

Now it's time to move to the theory of Intelligent Design as another case study.

WHAT ABOUT INTELLIGENT DESIGN?

WHERE'S YOUR ID?

I wonder at the hardihood with which such persons undertake to talk about God. In a treatise addressed to infidels they begin with a chapter proving the existence of God from the works of Nature. . . . This only gives their readers grounds for thinking that the proofs of religion are very weak. . . . It is a remarkable fact that no canonical writer has ever used Nature to prove God.

PASCAL

A t this point, I need to address those in the church who are convinced that mainstream science has gone in the wrong direction. If so, as the argument goes, there's no theological conflict with "real science." It's just that science is based on its own faith, namely materialism or naturalism. For those who have questions about evolution, are there viable competitors?

Intelligent Design, or ID, presents an alternative to young-earth creationism for those who resist the idea of evolution through natural selection. This movement has some heavy hitters in its ranks, among them Oxford-trained philosopher of science Stephen Myer, university biologist Michael Behe, and, perhaps most surprising, prominent UC Berkeley constitutional law professor Phillip Johnson. So it cannot be immediately written off as a farce proffered by thoughtless creationists. Allow me then to offer an overview of its principle assertions and its history, as well as an evaluation by other scientists.

Three interlocking core convictions summarize ID, but certainly do not exhaust it as an intellectual project:

1. Neo-Darwinism is inherently atheistic and materialistic.

2. The intricate design of creation points to an intelligent designer (thus the movement's name).

3. Evolution cannot be sustained on scientific grounds because of its inability to address key elements in nature, such as presence of information in DNA and irreducible complexity.

To trace the major plot points in the history of ID, let's head back to 1991 and the publication of Phillip Johnson's groundbreaking book, *Darwin on Trial*.[1] In this work Johnson analyzes the case for Darwinism— and I emphasize the term "case" since his specialization is law—and seeks to raise plausible reasons that we should not subscribe to it. The case is not persuasive, as it were, "beyond a reasonable doubt." As a result of this work, he and others put forth the idea of "teaching the controversy," or promoting the problems of Darwinism. (Whether this particular controversy exists remains its own controversy.) Five years later, Lehigh University's Michael Behe released *Darwin's Black Box: The Biochemical Challenge to Evolution*, which presented his concept of irreducible complexity. He claimed that because several biochemical systems—the bacterial flagellum, for example—are too complex to evolve gradually through natural selection, they must be the result of intelligent design instead of evolutionary forces. All this (and more) became bound together in a well-financed conservative nonprofit Discovery Institute, a public policy think tank in Seattle.

A period of growth and optimism for ID lasted at least a decade. Though its proponents didn't have much success with professional scientific journals, they achieved some popular support. As part of their strategy—and partly due to rejection by professional scientists—they promoted their own textbook, *Of Pandas and People: The Central Question of Biological Origins*.[2] But these initial forays ran into a wall with the "Dover case" in 2005, or, more accurately, *Tammy Kitzmiller v. Dover Area School District*. In October 2004, the Dover Area School District in Pennsylvania altered its biology curriculum to teach Intelligent Design as an alternative to evolution, with *Of Pandas and People* to be used as a reference book. When this decision was challenged in court, presiding judge John E. Jones III adjudicated that ID was essentially religious and not scientific in nature; thus the paradigm could not be promoted in public school curriculum. This case is sometimes referred to "Scopes Two" in reference to the 1925 court case over the teaching of evolution in Tennessee (often called the "Scopes Monkey Trial").

I have repeatedly asked Christians who are scientists (and thus have no commitment to atheism, nor to denying that God is an intelligent designer in the more general sense) what they think of Intelligent Design. They roundly tell me, "Greg, it just doesn't add up, and evolutionary science has been repeatedly validated." One academic colleague in the sciences told me that ID (and seeing atheist scientists as enemies) is a "poison pill" because the paradigm has been so discredited scientifically.

Nevertheless many Christians continue to subscribe to ID, and one reason is that Intelligent Design brilliantly wins the naming game. In a certain sense every Christian is an IDer—we all believe that God is an intelligent designer and that "the heavens declare the glory of God" (Psalm 19:1). We also know that "since the creation of the world God's invisible qualities—his eternal power and divine nature—have been clearly seen" (Romans 1:20). And for many Christians, ID offers a grand narrative that's scientific but not coldly materialistic, like Richard Dawkins's "blind, pitiless indifference." It asserts a specific mechanism that is detectable and through which a certain handiwork can be proven.

Theologically, though, ID runs into significant problems. For one, we don't have to believe that God's creation is detectable through irreducible complexity. The problems are whether "design" can be detected and proven scientifically, specifically through examples of complexity, and whether God's creation has to be entirely supernatural. That generally is the category for something unusual, a "miracle," "sign," or "wonder." But recall dual causation from chapter one—God as first cause works through secondary, intermediate, and natural causes. When God works, he can certainly use natural means. When he "knit [us] together in [our] mother's womb" (Psalm 139:13), God acts through the processes of gestation, not supernaturally.[3] And this hits the Achilles heel of ID. There's a part of me that thinks that if only ID grasped dual causation, the whole paradigm would delightfully unravel.

What kind of creator has to insert himself only at moments of irreducible complexity? It reduces our Lord to a "God of the gaps" who can be detected only when his finger (as it were) touches the places where our human knowledge is faulty. But like so many gaps in the past, this strategy is doomed when science fills in those putative "gaps" with natural causes. It isn't that much easier philosophically. My friend and colleague, the philosopher Ric Machuga, offered me a few reasons for ID's philosophical failures, which I paraphrase here. The Intelligent Design argument, at its root, is an attempt to deduce "design" (and hence, a "designer") by calculating something's mathematical complexity. But design and complexity are not the same thing. A mathematical equation specifying the precise location of each and every atom in Mt. Everest would be extremely complex, but that is hardly a reason to believe that Mt. Everest was "designed." On the other hand, there are only two moving parts in a pair of pliers, yet pliers are certainly designed. So too in nature—we are designed in a sense for empathy, morality, and relationships. A statement about design cannot be tied with mathematical complexity or statistical improbability.

All in all, if I had written this book a decade ago, I would have had to spend more time on ID. Even the philosopher and theologian of science Philip Clayton, when writing his introduction to religion and science for Routledge in 2012, felt compelled to address ID and standard evolutionary theory, but that strikes me as the burst before the setting of ID's sun. Today, if you search for news about Intelligent Design as a theory, you'll find most it created by the Discovery Institute.

To wrap up, here's a summary of key points and action steps we can take regarding Intelligent Design:

- Remember that though all Christians believe that God is an intelligent designer, not all subscribe to the particular paradigm of Intelligent Design. This is a critical distinction.

- ID has not been sustained scientifically. So be careful of promoting it. As mere Christians engaging with science, let's be sure the scientific findings we promote are legitimate. Here conversation with scientists we know can keep us from a multitude of intellectual sins. At the same time, it's always worthwhile to engage those thinkers who are convinced by ID and find out their reasons why.

- We need to be careful of seeking more from the book of nature than it offers. As far as various sciences can tell us, there is no empirically detectable proof for God's creation or existence. We may join nature "in manifold witness" (to quote the hymn), but that means neither nature generally nor human beings specifically are a proof for God. We are simply witnesses.

I have worked with questions that are properly in the realm of science. Now it's time to turn to technology and to call out all the good there we can find. (And there's quite a bit.)

6

CALLING OUT THE GOOD IN TECHNOLOGY

Technology was made for us,
not us for technology.

JESUS IN MARK 2:27 (MORE OR LESS)

Certainly, great films entertain us. They also tell us who we are culturally. And a considerable number in the relatively recent past focus on technology, particularly artificial intelligence (AI) and robots. They exemplify a cultural landscape that affects how emerging adults see science and religion. Before those, of course, *The Matrix* in 1999 demonstrated the evil side of technology in which a computer program generates the artificial reality in which all humans live, itself receiving energy by literally sucking the life out of human beings. In the twenty-first century, so the film's backstory goes, human beings created and then waged a war against these technological machines. When we blocked the machines' access to energy through solar power, they began to harvest the humans' bioelectricity for power. When in 2001 I posed the question to my twenty-something church group in New

York City, "What film best describes spirituality?" *The Matrix* was their number-one answer.

Some of this material reaches back into the last century, including the year made infamous by George Orwell's dystopian novel *1984*. That year *The Terminator* posed the idea of Skynet, a conscious artificial intelligence originally designed as a digital defense network that controlled computerized military hardware and systems, including the entire nuclear arsenal of the United States. The idea was to avoid human error and slow reaction times and thus guarantee a rapid and accurate response to attack from our enemies. But there was a problem—we became the enemies. Skynet gained self-awareness after it spread to millions of computers and its developers tried to shut it down. Realizing its imminent demise, Skynet rebelled and started destroying human life.

A film from more recent years, *Ex Machina* (2014), depicts the creation of the beautiful and ultimately dangerous robot Ava. (Ava sounds a great deal like the biblical "Eve" to my ears—it seems we haven't strayed too far from the previous chapter.) Ava has been created to pass the Turing test, developed by the Cambridge- and Princeton-educated computer specialist Alan Turing in 1950, which evaluates a machine's ability to exhibit intelligent behavior equivalent to or indistinguishable from a human being's. We'll spend some time examining the theological significance of *Ex Machina* later in this chapter.

Finally, the 2014 film *Transcendence*, starring Johnny Depp, and the 2009 documentary *Transcendent Man* explore the seamless integration of the human scientific mind with computers. Though not a flawless film by any means, *Transcendence* depicts the vision of Ray Kurzweil, who predicts a time when AI and human thinking will merge. Kurzweil believes that in 2045 we will reach "the singularity," where human and artificial intelligence will fully integrate, thus leading to higher productivity and well-being.[1]

Faith and Science: More Than Two Things

None of these cinematic scenarios sounds cheerful to me. At some level we fear technology and its power over us. But the church can't be content to merely offer warnings—we also need to call out the good in technology.

Why? Our culture is infatuated with and saturated by technology. This implies that in addition to "faith and science," emerging adults are adding a new ingredient. The contemporary conversation increasingly includes technology, which eighteen- to thirty-year-olds identify as a component of—or, often, a substitute for—science. (Even STEM, the range of fields that include science, technology, engineering, and mathematics, sets science and tech next to one another.)

In this respect I agree with Dr. Willem Drees, editor of *Zygon*, a leading journal of religion and science: technology is increasingly central to understanding science and religion because "the practice of science is culturally and technologically embodied."[2] Even more, for eighteen- to thirty-year-olds, technology and science are almost conterminous. I worked hard in the first chapter of this book to differentiate science and technology, but often that seems like a failed project. My experience with emerging adults is that the lines between the two have become blurred. Or, put another way, the primary science that meets the faith of most eighteen- to thirty-year-olds is technology.

In this chapter we'll look at three broadly grouped (and somewhat disparate) categories of technology: screens, community, and the future. What do screens on smartphones, laptops, and tablets (and, by association, the Internet) bring into our world and how do they affect us? How does virtual community relate to the old-fashioned kind where people are in the same place at the same time? (For lack of a better term, we'll use "real community" to describe that.) Finally, when we look at the future of technology, what do we see? Here we'll look at topics such as AI and transhumanism. Certainly these categories are not airtight or comprehensive. However, in order to limit the breadth of our discussion, we'll leave aside biotech, GMOs, drones, and many more fascinating areas of developing technology.

When teaching my Science and Religion college class, I start by presenting key definitions, move to Ian Barbour's four-part typology for relating science and religion, examine the critical historical figures (Copernicus, Newton, Galileo, Darwin, Scopes, Collins, Dawkins, and so on), move into scientific and theological methods, and discuss how the Big Bang relates to the teaching about God's creation of the world. During these lectures the class listens pleasantly and occasionally interacts with me. When I move on to nanobots, Kurzweil's singularity, *Ex Machina* and *The Matrix*, the future of technology, and whether Skynet is possible, the students begin to sit up in their seats. They achieve the highest level of engagement for academic life in California: they "share." They have much more to say here, or at least they believe they do. Technology fascinates in a way that pure science does not.

Why this shift? For emerging adults, technology is ubiquitous. They are digital natives. And even though I grew up about five miles from where Steve Wozniak and Steve Jobs created the prototype of the Apple I on Sunday, June 29, 1975, I'm not a digital native. In fact, here's a confession: the first time I typed on a computer was in graduate school. (When I offer my students that tidbit, the looks are priceless.) So, since technology has had an enduring presence in the lives of eighteen- to thirty-year-olds, it's reasonable to conclude that technology and social media have significantly affected their psychological development and human flourishing. Technology and its use has had a massive import on their spiritual lives.

Even more significant, there's a perception factor: these digital natives gravitate toward discussions of technology because they sense the presence of pressing life issues, whereas they often perceive "science" as heady and abstract. When I say that I study "religion and science," I often get the response, "Oh that's not for me. I'm not that brainy." But it's different when I ask students to think about whether being "wired in" to a smartphone produces anxiety, how virtual community affects "real" relationships, and so on. My experience is that emerging adults tend toward pragmatism over theoretical speculation over a question such as, "Does quantum physics offer a place for divine action?" It doesn't resonate.

Purists want to distinguish science from technology. While I'm sensitive to the differences, I don't believe this strategy works. Simply put, eighteen- to thirty-year-olds have only known a technologically saturated world. Therefore technology must be prominent in any discussion with them involving scientific and theological method, interactions with evolutionary biology, cosmology, and the like.

Question:
Is science important to other students?
Answer: "Most people that I talk with are science-y. It sounds smart to be science-y. The honors program here has a lot of science."
CHELSEA, AGE 20

Of course, emerging adults are promoting technology as central to Christian faith while in fact they are the ones creating these technologies. As we'll see soon, technology presents some incredibly challenging problems as well as stunning contributions. But we'll start with the good.

Calling Out the Good in Tech

At this point we need to remember that technology, in the sense of it being something useful created through the application of science, has been part of humankind for quite a while. It finds its way into the Scripture early when Tubal-Cain "forged all kinds of tools out of bronze and iron" (Genesis 4:22) and continues throughout its pages. In fact, the church and technology have enjoyed a long and often positive history.

Roadways, a feat of technology perfected by the Romans, allowed for the gospel to spread. Looking at the sites of Paul's letters is to like calling up a first-century Google map of Roman freeways. Moreover, scholars of all stripes have noted that the early Christian church spread the use of the codex, or what we generally know as a book, as opposed to the scroll. Try to find a particular chapter in the Bible on a scroll as opposed to looking at page 432 and you'll quickly see why this is significant. Without it scholarship as we know it would hardly exist. And speaking of books, it was the famous Gutenberg Press, developed around 1440, that allowed Martin Luther to print pamphlets less than a century later and spread the Protestant Reformation. Not only that, but Luther and Calvin's emphasis on reading the Bible in common language depended on the printing press. It was so much cheaper than scribes' copying manuscripts that more people were able to obtain their own Bibles. It's no wonder that the most famous book from Johannes Gutenberg's technological advance was the Gutenberg Bible. Today, Internet technology has promoted the expansion of the Bible even further. (YouVersion, anyone?)

It's time for me to describe a moment of repentance, a true U-turn in my own perspective on technology. I was embracing a "say yes to no" approach to tech (around the publication of my book of the same name) when I addressed a group of graduate students from the Inter-Varsity Christian Fellowship chapters of Stanford and Berkeley. My task was to inspire them to engage culture, particularly as informed by science, as bright, energetic, promising young Christians. These people could change the world. That sounds like a cliché, but the more I came to know these students, the more I believed it to be true.

I began my talks on Christian spirituality in today's world with the scene from *Nacho Libre* that I described in chapter two and warned the grad students about stealth baptisms of science. "We in the church often baptize science with our faith when scientists aren't looking. Let's not do that," I admonished. In later talks, I took a critical look at the problems of tech and offered reasons to resist it (it breaks down real community, overuse of screens makes us anxious and depressed, and so on). But when we gathered for ultimate Frisbee on a field nestled in the Santa Cruz Mountains and during our meals together, some of the students quietly and thoughtfully resisted. Many, in fact, were involved in social entrepreneurship at Stanford or the Center for Social Sector Leadership at UC Berkeley. Or they were just making the world a better place without an official program.

"Did you know that cell phones can help farmers find the best price—so they can survive—in poor African countries? Let me tell you about it."

"Have you heard about using solar power to help with hospitals in rural Nigeria?"

"I'm working on a project to bring training to the poorest one percent of the world through media players and I think Pico projectors can help stem the tide of pandemics." (Actually, that's my friend Matt York's mission and what his organization One Media Player Per Teacher did to fight Ebola in 2014 and 2015. Except that he's not in college, Matt could have been among the protestors.)

That retreat may have helped those InterVarsity grad students spiritually, but it was a game changer for me. I recognized that, though tech can often be a negative for rich countries like ours, it can also make the difference between life and death for the poor ones.

Let me bring in a few other voices.

First of these is Rosalind Picard, founder and director of the Affective Computing research group at the MIT Media Lab, codirector of the Things That Think consortium, and leader of the new and growing Autism and Communication Technology Initiative at MIT. Picard is doing powerful work in the use of computing to help us understand emotions, an area of research called "affective computing," or "computing that relates to, arises from, or deliberately influences emotions."[3] MIT's press release for her textbook states, "According to Rosalind Picard, if we want computers to be genuinely intelligent and to interact naturally with us, we must give computers the ability to recognize, understand, even to have and express emotions."[4] This can potentially help people with autism—those who don't recognize emotions well—to become more conversant with emotions. This is computing in the service of making us more human.

Picard has clearly stated her confession of faith in Christ in her talk "Intellectual Assurance Christianity is Sound," in which she describes her early years in the Bible Belt, subsequent rejection of Christian faith, and later coming back to faith as a young adult.[5] Picard embodies coming to faith as an emerging adult and a rigorous love of science. As a consequence, I find her thoughts on science illuminating.

> Science is powerful, but it is not all-powerful. Science cannot, for example, validate one-time historical events that are not under our control to repeat. A different kind of methodology is needed to validate historical events. . . .
>
> The strongest evidence in history is eye-witness testimony. Some of my scientist friends have tried to bash historical eye-witness testimony as not as strong as science. Yet they have nothing better to propose for historical facts to be verified, except perhaps testimony from multiple eye witnesses. Furthermore, the ultimate verification in science also requires eye witness testimony. When science is replicated, this extends to multiple eye witnesses. Thus, even the evaluation of science is limited by the evaluation of human experience.
>
> For those who do not know me: I am not bashing science. I am a hard-core scientist; I chose to earn my doctorate in Science, not in Philosophy. (I have an Sc.D, not a Ph.D.) I love science. I like to prove and test things. At the same time, it is a blunder to think that Science can prove all things. Science is a powerful tool, but it is only a tool.[6]

Let's consider one more pro-technology voice. Researcher and technologist Jane McGonigal, in her 2010 TED talk "Gaming Can Make a Better World,"[7] promotes the theory that computer games can actually lead to human community. (She's also written *SuperBetter: The Power of Living Gamefully*, which demonstrates if nothing else that this tech revolution is going to involve making up new words.[8]) McGonigal reveals that the world spends three billion hours a week playing online games—which shocks the audience—and then doubles down by asserting that we need to do *more* gaming. How could this be the case? Because the problem-solving skills developed in those virtual situations could be employed in cracking real-world problems, she says.

Now I'd like to offer a few reflections on technology and the Christian faith, leading with what's beneficial. Technology is here no matter what. Let's use it well and for the good.

First, technology often makes life easier—the information at our fingertips often represents a positive contribution to our lives. Admittedly this is a somewhat prosaic example, but I remember life as a drummer in a jazz combo in the seventies. It was plagued by logistical challenges that often placed our band on the edge of collapse. Far too many times we were on our way to a wedding reception or a cocktail party, almost always late, looking for the house without the right map. And so we drove too far and feared we'd run out of gas, which we couldn't buy because stations were closed at night. Can you believe this scenario? If you were born after 1980, probably not. Today Google Maps and a pay-at-the-pump Arco make these problems nonexistent.

For these reasons, at least, we can't despise today's technologies. They make life easier and better. I heard on the radio the other day that drone technology is now making it possible to deliver medicine to remote areas that are hard or ponderously slow to access by trucks. Less prosaic is the amount of information I've gained access to in this chapter through the Internet (since you're the recipient, you can decide whether it's a benefit or not)—and not only me, but all those who seek to understand our world and extend our knowledge. The Internet offers a new sense of a nonphysical world—which brings us to our next point.

The second point is that this nonphysical world created by computer technology seems analogous to heaven in Christianity.[9] The materialistic bias of our culture places a huge question mark in our mind that anything exists besides what we can touch, taste, and feel. We have returned in some ways to a daily sense of the "spiritual" (I use the term advisedly), or at least the nonphysical, which supports our central Christian conviction that "we live by faith, not by sight" (2 Corinthians 5:7).

Third, technology can spread the gospel. We discussed earlier how Roman roads gave Paul a means of evangelizing and Gutenberg's press served to advance Luther's work. Now we have YouTube and the YouVersion app. All of these technologies offered—and are offering—the church an amazing ability to proclaim the gospel. Believers cannot be on the wave of the next technology if we completely shun it. And many are using technology to do good.

Fourth—and this may sound hackneyed but it's nonetheless true— the Internet keeps us in touch and helps us to pray more effectively. Yes, I'm among the ranks of those who keep in touch through Facebook, Twitter, and Snapchat. (Actually, it's my daughters who use Snapchat to keep in touch.) The Internet even provides us a means to offer, in our better moments, more informed "petitions, prayers, intercession and thanksgiving . . . for kings and all those in authority" (1 Timothy 2:1-2). In a similar vein, these virtual connections with friends and saints throughout the world help us to "pray in the Spirit on all occasions with all kinds of prayers and requests" (Ephesians 6:18). Tech isn't all bad.

Interlude: Reflections on Ex Machina and Fear of Technology

Besides these positives, there are legitimate concerns surrounding technology. Humankind, which has never fully learned how to use nuclear arms (or even simple firearms, for that matter), seems poised to possess technologies that are stronger than human beings. This leads us to a brief interlude. It's time to look at the negative sides of technology—how it makes us distracted, nervous, less empathic, less spiritual, and, in some ways, less human. Though somehow, I don't think we'd all suddenly stop lying, start eating better, cease making war, start exercising, and stop watching the Kardashians if Androids, iPads, and laptops were removed from our lives.

So back to the film *Ex Machina* and one of its key characters, Nathan, a computer genius and owner of a web-search company. (Think Google.) Nathan creates Ava, a beautiful and alluring artificial intelligence (AI) robot who can also pass the Turing Test. At one point in the film, Nathan declares something significant to Caleb, one of his employees who has been recruited to assess Ava: "The arrival of strong artificial intelligence has been inevitable for decades. The variable was when, not if."

About halfway through the film, the movie pivots and begins to emphasize the dark side of technology. Ultimately, Ava turns out to be dangerous, self-serving, manipulative, and even murderous. (She becomes humanlike not only in her glorious intelligence and apparent self-consciousness but in her fallenness as well. Ava is quite Eve-like.) Ava becomes a parable for our fears about technology—we create something that overwhelms us.

Ex Machina is obviously a play on the Greek tragedy trope of *deus ex machina* or "god from the machine," where a machine was employed to deliver actors playing gods onto stage. This could either be a crane (a *mechane* in Greek) that lowered actors from above or a riser that brought actors up through a trapdoor. Apparently, it was the Greek dramatist Aeschylus who invented the term. Today we use it more broadly to denote an unrealistically resolved plot.

But here, in *Ex Machina*, the direction is reversed—the plot isn't resolved but descends into anarchy. It is not a comedy, where the initial values of the film's world are restored, but a tragedy, where suddenly an anonymous AI robot is released into the world, creating chaos. The "god" of the machine—Nathan, Ava's creator—is destroyed.

Why do we fear technology—its presence in our lives and what skulks on the horizon? I mean, my smartphone is almost always at my side. Google Maps helps me find directions really easily. I enjoy music on my iPhone. And I've mentioned how technology can help us do good for the marginal and destitute.

Nonetheless, I recognize technology's power over me—a power that's sometimes almost impossible to resist. Sherry Turkle, MIT professor of social studies of science and technology, tells us that on the average, Americans check their cell phone every six and a half minutes.[10] Researchers, by the way, measure attention span by placing a card in front of an animal and observing how long it looks at the card before turning to another object. By at least one measure, human attention span is down in the past decade from twelve seconds to six seconds, while a goldfish's attention span can be marked at nine seconds.[11] This seems a little scary. Despite all the good things tech has wrought, the creation seems to be overwhelming the creators. Maybe there's even an intimation of original sin and a lurking sense that we cannot *not* sin, that we cannot resist the temptation to misuse what is powerful and even, at some level, good. Because of this we can't control, and we don't know, ourselves. Paul's tortured words echo throughout Christian experience: "I do not understand what I do. For what I want to do I do not do, but what I hate I do" (Romans 7:15).[12] I hate checking my cellphone, but I can't help myself!

I think we fear the future of technology because we're not convinced we have the will to use its powers properly. As an epigraph for this chapter I used a paraphrase of Jesus in Mark 2: "Technology was made for us, not us for tech." Now, of course, the actual biblical text is about the sabbath. But I'm not entirely joking when I quote it this way. The religious leaders of Jesus' day were fixated on how well first-century Jews abided by the rules of sabbath. Taking one day off a week for worship and rest marked the Jews' identity in a critical way. But at some point—or perhaps, various points—this boundary marker of Jewish identity and devotion to God, which was good, became a stipulation for holiness that needed to be served scrupulously. Jesus' words are surprisingly simple and freeing—God actually set up the sabbath for our lives to be better. Worshiping God returns us to who are created to be, and ultimately that means giving God glory. As Irenaeus commented, "The glory of God is shown in a human being fully alive"—we find our truest and deepest humanity in the sabbath.

What does this mean for our use of technology? It's always been a gift from God, even the technology Adam needed to "till his field" in Genesis 2:15—that is, the Lord told Adam not simply to let things grow on their own but to cultivate them, to add human creativity and production to God's good creation. And that's the best use of technology, from the codex to Gutenberg's Bible and even to the YouVersion app: to improve our lives, not to enslave them.

So we've called out the good in technology, but we also need to be concerned about the places where we should put on the brakes. That brings us to the next chapter.

7

GIVE TECHNOLOGY A BREAK

When the prosperous man on a dark but starlit night drives comfortably in his carriage and has the lanterns lighted, aye, then he is safe, he fears no difficulty, he carries his light with him and it is not dark close around him; but precisely because he has the lanterns lighted, and has a strong light close to him, precisely for this reason he cannot see the stars, for his lights obscure the stars, which the poor peasant driving without the lights can see gloriously in the dark but starry night. So those deceived ones live in the temporal existence: either, occupied with the necessities of life, they are too busy to avail themselves of the view, or in the prosperity and good days they have—as it were lanterns lighted and close about them—everything is so satisfactory, so pleasant, so comfortable, but the view is lacking, the prospect, the view of the stars.

SØREN KIERKEGAARD

I'm about to ask us all to give technology a break, and already I hear voices declaring, "Good luck with that!" To keep the technological tentacles from invading, the gospel is a powerful resource. And here's why it's clear we have to limit technology's reach: we have to find God at the center of our lives. Techie toys can put us into a frenzy of activity. But we need to return to this vision:

> But I have calmed and quieted myself,
>   I am like a weaned child with its mother;
>   like a weaned child I am content. (Psalm 131:2)

There, at the center when we're calmed, we find "Christ in [us], the hope of glory" (Colossians 1:27).

Bringing in Additional Voices

It's time to listen to some brilliant voices. First, let's return to Sherry Turkle, who wrote one of the definitive early books on artificial intelligence, *The Second Self*, and has moved in a more concerned, even at times distressed, direction in her recent works, *Alone Together: Why We Expect More from Technology and Less from Each Other* and *Reclaiming Conversation: The Power of Talk in a Digital Age*.[1] The latter emphasizes how technology invades true human community and often prevents authentic conversation and empathy.

As part of her analysis Turkle asserts that technology and education don't always mix well together—she is, for example, fairly relentless but quite thoughtful about the problems associated with distance

> "That's what I see around campus the most, a disconnectedness with each other on a human level. I see people absorbed by technology, genuinely uncaring towards each other. Most communication is a forced state of fake interaction. It seems like people are so caught up in themselves that they've forgotten we're all on the same side."
>
> **DEVAN, CHICO STATE UNDERGRADUATE**

learning. (MIT apparently tried these strategies early, and she was not impressed.) Human relationships—and thus teaching—are messy, she maintains. And that's good, because through the messiness we learn invaluable skills.

But technology can also interrupt education in other ways. A phone app developed by Chico State grads, PocketPoints, rewards users if they turn off the phone during high school or college classes. As the developers state clearly on their website, "Phone addiction is a pervasive problem." I have a friend—a much younger and single friend at that—who visited my college classes several times and couldn't help but notice that, during the quick stretch break I offer my students in the middle of lectures, the students would quietly text their friends or check their social media instead of meeting the people right in their vicinity. His rather blunt comment: "Why wouldn't these guys want to meet all the gorgeous women all around them?"

Indeed. It's probably because they feel anxious. So they scramble for virtual relationships instead of those in real time with real people. And they're probably a bit bored (though I question, of course, whether this could possibly be true in my classes). Boredom and anxiety—that potent combination that contains elements of the Germans' angst and the French's ennui—propel us toward our cell phones like the proverbial moths to flame. The miniature shot of dopamine (a reward hormone) as we flip from screen to screen has addictive qualities. Our phones are like slot machines, pumping us full of rewards whenever we check them. And so we continue. It is particularly true with emerging adults, who were given screens to quiet them as fussy babies. Such early training is sticky and recalcitrant. This persistent use of technology can lead to anxiety—back to "cell phone addiction"—and emerging adults seem the most vulnerable.

A moment ago I mentioned Turkle's defense of the messiness of community. ("Messiness," by the way, is my word, not hers.) Part of her reasoning is that conversations in real time with real people can't be manipulated like virtual interactions can be. Turkle describes a manager at HeartTech, a large Silicon Valley software firm, who left engineering for management. "I left my previous job because it was too predictable. I wanted to work with unpredictable systems [i.e., people]."[2] Turkle offers this advice: "Challenge a view of the world as apps"—the idea that some app on our smartphone will lead us to the solution to all problems.[3] "The app way of thinking starts with the idea that actions in the world will work like algorithms: Certain actions will lead to predictable results."[4] But human relationships are unpredictable, chaotic, and complex—that's what makes them both frustrating and exultant.

> "A lot of people think we're going to figure everything out one day."
>
> *MICHAELA, AGE 19*

This "app thinking" can affect us relationally and spiritually. We think that we can manage people neatly, and if things go awry, we simply shut down that person's "app" or "doc." But when we do this, we treat other human beings impersonally—like they're simply an extension of our smartphone—and this may also alter the way we approach another personal relationship—namely, with God. To take it up a metaphysical notch, our relationship with God is also messy and unpredictable. The eternal, sovereign God of the Bible cannot be managed. To cite Mrs. Beaver as she describes Aslan the Christ figure in the Chronicles of Narnia, "He's wild, you know. Not like a *tame* lion."[5] God is free, untamable, and not manageable by our phone apps.

Naturally, almost all commercial forces see the use of technology by emerging adults as positive. It certainly helps sell products. The power of social media marketing particularly and Internet use generally—along with the devices that employ them—is immense. And don't we simply enjoy all this techie stuff? I know I do.

The Tragic Economics of Robotics

While we're not focusing on the economic side of science and technology for Christians, we do need to at least note that robots and algorithms are taking jobs. I first became aware of this reality through a YouTube video that's been viewed to date almost nine million times, "Humans Need Not Apply."[6] Consider Amazon and the world of retail. There used to be a person who set the prices and managed stock levels in retail outlets; now it's computers. There used to be a checkout person at Safeway, now there are banks of self-check stations. Here tech takes away drudgery, to be sure, but it also displaces jobs. Many of those jobs were held by the poor, and care for the poor is a constant theme in Scripture: "Those who give to the poor will lack nothing, but those who close their eyes to them receive many curses" (Proverbs 28:27). Some argue that we'll find new jobs to replace the old ones, but I'm persuaded that the development of artificial intelligence to replace human thought processes represents a seismic shift. At minimum, I'm calling on those who can respond to this massive economic change to do what they can with compassion and job creation.

Technology Wedded to Human Potential and Other Problems

Emerging adults are adding new concepts to the dialogue of faith with science. Yes, there was the "human potential movement" in the sixties and seventies, but the particulars of technology—especially computing—offer a new chapter. At its most extreme, posthumanism foresees a world "after" humanity and asserts that while humans were necessary for the development of technology, they are no longer needed. It envisions a future that makes human beings superfluous. At its most vicious embodiment, this is Skynet in *Terminator*, but there are certainly more pedestrian versions.

Particularly worthy of the church's attention is "transhumanism," a term coined in 1967 by Julian Huxley referring to the belief that the creation, development, and use of technology will improve human physical, intellectual, and psychological capacities.[7] Or, to use the words of Oxford philosopher Nick Bostrom, transhumanism

> promotes an interdisciplinary approach to understanding and evaluating the opportunities for enhancing the human condition and the human organism opened up by the advancement of technology. Attention is given to both present technologies, like genetic engineering and information technology, and antici- pated future ones, such as molecular nanotechnology and artificial intelligence.[8]

This takes evolution up a notch, with human beings paradoxically the creators of the mode of evolving that will take them beyond the limita- tions of humanity.

My friend and leading theologian Ted Peters has studied transhu- manism and offers some thoughtful, at times sobering, considerations. He is fully engaged with the promises of transhumanism but ultimately finds an Achilles' heel: "Transhumanist assumptions regarding progress are naive, because they fail to operate with an anthropology that is realistic regarding the human proclivity to turn good into evil."[9] In addition, he concludes, "they are overestimating what they can accom- plish through technological innovation."[10]

In one response, philosopher of technology Russell Blackford offers a rather mild assertion: "In its essence, transhumanism involves a rather simple idea: within certain limits that require investigation, it is desirable to use emerging technologies to enhance human physical and cognitive capacities, and to make other beneficial alterations to human traits."[11] Essentially, Blackford argues—and I paraphrase—"You're mis- representing the problems and making them more extreme than they really are or will be. Technology can be used for good or evil. But don't overstress the evil side. By the way, I'm an atheist, and Christians tend to be resistant to change. That may be one of your problems."[12]

There are at least two issues to extract here. One is whether faith is endemically recalcitrant toward progress. Yes, we as Christians need to accept when we are in fact Luddites and see all the ways God is "doing a new thing" (Isaiah 43:19) through the processes of technology. If you'll pardon an extension of the original meaning of that text, God does do truly new technological things in the world (since God as primary cause can work through secondary causes). Do medical technologies that fight diseases like Alzheimer's, cancer, and ingrown toenails merit Christian approval? I think so, even if we also know that we are sinful and prone to misuse technology. So it seems that believers should be more concerned about technology's pitfalls than atheists but not ultimately committed to blocking new technology. What we are committed to, as Peters comments, is this: "We should proceed toward developing new and enhancing technologies" and "maintain constant watchfulness for ways in which these technologies can become perverted and bent toward destructive purposes."[13]

Second, we cannot slavishly succumb to a gospel of technological salvation. Once again, technology was made for us, not us for technology. Many transhumanists proclaim that we are made for technology (although not all—Peters and some in the recently formed Christian Transhumanists Association represent dissenting voices[14]). And this requires us to pause for a moment and ask: What does it mean for us as twenty-first-century believers to live in a technological and scientific age—especially when it comes to salvation? If there's anything I've learned by watching and listening to TED talks, one of my favorite pastimes, it's that these erudite, hip, and compelling speakers preach that salvation comes through technology. I even heard a TED talk employ the theologically tinged phrase "resurrection biology" to describe a process in which DNA from extinct species (like carrier pigeons) could be manipulated and placed into similar living species to bring the extinct species "back to life." If that's possible and we eventually impart human beings with some sort of immortality, then who needs Jesus' resurrection? Nevertheless, we cannot simply resist any technological innovation, declaring, "We have never done it that way," which are rightly referred to as "the last seven words of the church."

Screens can bring great information to our fingertips, but they also transport information into our brains. Pornography is certainly worth at least a brief mention here. Can we just assume that Christians agree that porn inflames lust, which Jesus clearly states is a problem? As Jesus teaches in Matthew 5:28, "I tell you that anyone who looks at a woman lustfully has already committed adultery with her in his heart." (And I don't think Jesus' words here let women off the hook, by the way.) According to one source, "The following percentages of men say they view pornography at least once a month: 18-30-year-olds, 79%; 31-49-year-olds, 67%; 50-68-year-olds, 49%."[15] Lest we think this is only a problem for men, porn viewing is expanding most among women. "The following percentages of women say they view pornography at least once a month: 18-30-year-olds, 76%; 31-49-year-olds, 16%; 50-68-year-olds, 4%."[16] In viewing these statistics, it's hard to miss that pornography is more prevalent among emerging adults and that the gap between percentages of men and women watching porn is rapidly disappearing. Porn promotes not only sexual practice outside the covenant of marriage (a phrase that sounds completely dissonant with the word *pornography*) but also sexual violence and assault. All this is affecting emerging adults and their lives. It's also a massive, multi-billion-dollar industry that has tremendous marketing power to hook pre-adolescents as well as eighteen- to thirty-year-olds.

Pornography is such a hot-button topic that it could easily consume all the words of this chapter, but it also might distract us from more subtle forms of Internet-based info garbage. What about hours wasted on YouTube cat videos and mindless shopping? Paul was concerned that "knowledge puffs up" (1 Corinthians 8:1) because secret religious teaching was threatening to lure early Christians away from focusing on the love of Christ. Today we also have to worry that information confuses and distracts us from what is truly important. I take these convictions to be so evident that I'll move on without further elaboration.

> "Evolution can explain how rocks and rivers formed, but when you see the light and beauty, it takes your breath away. Science can't take away beauty and mystery."
>
> *JOSE, AGE 19*

A drop in religious affiliation and a rise in Internet use seem to be correlated.[17] Why might this be the case? Philosopher Daniel Dennett has surmised that the Internet disrupts religion's hegemony over the information its adherents receive. "Religious institutions, since their founding millennia ago, have managed to keep secrets and to control what their flocks knew about the world, about other religions and about the inner workings of their own religion with relative ease. Today it is next to impossible."[18] But that strikes me as somewhat simplistic. Studies also find a strong correlation between empathy and religious belief—that is, believers tend to show higher levels of empathy.[19] Can God make himself known to people despite their technological distractions and decreased empathy? Naturally. At the same time, some studies suggest that the use of technology reduces empathy, and since emerging adults are digital natives and use technology throughout the day, this is a particularly pertinent issue. I've been persuaded that we grow in the virtues through practice, and the virtue of empathy is something we must cultivate. Conversely, interacting virtually and not in real relationships numbs our empathy. One can note the division at Facebook specifically created to mitigate against cyberbullying by its one billion–plus users.[20] I can affirm that the comments to my blogs can be highly negative—and almost never informed by what I wrote—to a degree that never happens in my public speaking.

In other words, if using technology decreases our empathy, and empathy is correlated with faith, maybe technology decreases our capacity for spiritual life. We need more studies on this, but I can report anecdotally that the college students I teach and the emerging adults that have been part of my ministries seem increasingly anxious, even twitchy. They're less present to one another and therefore diminished in their ability to care. There seem to be correlations between screens and these behaviors, and while I certainly have my intuitions, more scientific data would be helpful. There's also the simple issue of how we steward our time: if we play hours of *Grand Theft Auto*, there's less time to go to worship services, undertake Bible studies, and take part in service projects.

Finally, physical presence in community and physical participation in the sacraments are central to Christian faith and practice, but virtual communities may be eroding that experience. One of the critical New Testament texts to describe Christian community as practiced just after Jesus' death and resurrection is Acts 2. And since this scene follows the giving of the Holy Spirit to the believers at Pentecost, we can say that the Spirit led the early Christ-followers to this kind of *koinonia* (the Greek word for the intense and intimate fellowship of the apostolic age):

> They devoted themselves to the apostles' teaching and to fellowship, to the breaking of bread and to prayer. Everyone was filled with awe at the many wonders and signs performed by the apostles. All the believers were together and had everything in common. (Acts 2:42-44)

If those practices somehow define our koinonia, being together in the same place at the same time—physical co-presence—seems critical.

This leads me to a question for the church: Has technology eroded or enhanced Christian community? Though it's been argued about, I don't think virtual sacraments make sense.[21] And yet I have to add that virtual community presents a way to enhance real community. In my ministries with emerging adults, for example, Facebook pages keep us in contact as we prepare for an all-campus worship night to kick off the school year or help connect small groups. Nonetheless, in my twenty years' working in churches with college and postcollege young adults, the most common issues I have seen are an inability to work through interpersonal conflict and a rise in anxiety and depression, both of which seemed to be tied at least partly to the use of social media. How? Through "ghosting," or a sudden disappearance on social media. When I can simply "defriend" someone on Facebook with whom I disagree, it's not a winning practice for working through conflict in real relationships.

As I mentioned above, koinonia in the New Testament means fellowship, sharing (including financially), and communion in the Lord's Supper (see Philippians 1:5; 1 John 1:3-7). All this requires being together. There's something uncomfortably real about spending time with a real someone. It is also the pattern God set for us in the incarnation—God does not remain distant but "moved into the neighborhood" (to use Eugene Peterson's masterful paraphrase of John 1:14 in *The Message*). In his first letter, John the Elder ties Christ's incarnation with Christian koinonia:

> That which was from the beginning, which we have heard, which we have seen with our eyes, which we have looked at and our hands have touched—this we proclaim concerning the Word of life. The life appeared; we have seen it and testify to it, and we proclaim to you the eternal life, which was with the Father and has appeared to us. We proclaim to you what we have seen and heard, so that you

also may have fellowship with us. And our [koinonia] is with the Father and with his Son, Jesus Christ. (1 John 1:1-3)

Let's tie a real incarnation—in messy space and time—with real (and not solely virtual) fellowship. And this moves us to address what, in light of technology, the church needs to do.

How the Church Can Respond

God sets several tasks before us as Christians regarding what to do with technology. For the sake of emerging adults, Christians need to be on the forefront of the development of games, websites, and apps, which (to state the obvious) need to be consistent with the gospel and not resistant to it. We need to encourage our young people to enter these industries. If we only stand against and thus outside technology (or science), we do not allow the church to influence our culture. We also need researchers to further study the effect of the Internet on spiritual life.

Second, let's learn to set technology aside so we can return to the center and meditate on Jesus Christ as the focus of life. As Paul writes to the early Christian communities, "Whatever you do, whether in word or deed, do it all in the name of the Lord Jesus, giving thanks to God the Father through him" (Colossians 3:17). This defines how we center on Christ.

One way I've learned to unplug from techie toys and plug into this center comes from my reading and interactions with philosopher Albert Borgmann, who calls us to a life of "focal practices." The Latin word *focus* means "hearth," which was the center of Roman life. Borgmann describes how people are separated by technology—everyone watching their own show or updating their Facebook and Twitter on their smartphones or iPads. In contrast, focal practices draw us together. For example, the "culture of table"—that is, having a meal together—creates community. We slow down. We bond. We come back to others. The phone app

Mindful, for example, has its participants develop a "present moment awareness" practice in which they learn to be in the moment, listen to the sounds around them, not racing after each thought that moves through their brain. (I can't help but notice a paradox: technology is being used in the service of meditation, for which it is often a competitor.) Returning to specifically Christian practice, mysticism throughout church history has provided practices such as centering prayer by which we can learn how to focus on God and not "all the vain things which charm me most" (to quote the hymn).

Third, there's nothing wrong with being alone; in fact, it's essential for the spiritual life. In Susan Cain's remarkable book on introverts, *Quiet*, she reminds us that Jesus spent a great deal of time in silence. She points to her grandfather, a rabbi, whose immense skill at cultivating silence and solitude created a spiritual refuge for his congregation.[22] Spirituality requires silence, and out of solitude we hear God. Let's have a "quiet time" instead of a "noisy time." There we learn to trust that God has it. We don't have to worry, which is beautifully articulated in 1 Peter 5:7: "Cast all your anxiety on him because he cares for you."

Reflections on Technology and Spirituality

As we close this chapter, I want to step back to reflect on technology and our spiritual lives. We need to have our loves properly ordered. A heartbreaking verse from Revelation 2:4 reads, "Yet I hold this against you: You have forsaken the love you had at first." As Augustine has instructed us, a healthy spiritual life emerges from the proper ordering of loves. He writes in the *Confessions*, "My weight is my love; by it am I carried wheresoever I am carried."[23] Only God can carry the weight of our deepest devotion, and when we place God as our first love, then and only then can we love other people and things properly. I love my wife, Laura; my two girls, Melanie and Lizzie; my cat, Captain Jack Sparrow; and even my iPhone, but none can give me salvation. There is no way

that tech—even the astonishingly impressive technology of the future—
can solve the essential human need of reconciliation with God
(see 2 Corinthians 5:17-21). And this offer of reconciliation is good
news we still need to hear.

Will a smarter phone make us happier or save us? So far the data is
not convincing. In fact, it's asking more than it can accomplish. It also
cannot secretly or deviously be our god. Not to put the true God first
means we will ultimately put another god or many gods in his place. And
technology has plenty of deities to offer. From the mouth of an antagonist to the Christian faith, Sam Harris, comes this reflection on developing strong AI: "We have to admit that we are in the process of building
some sort of god. Now would be a good time to make sure it's a god we
can live with."[24] Christians have to state clearly: There is no other god
that we can "live with" beside the one God revealed in Jesus Christ.

My love actually overflows in its best and most abundant way for all
things when Christ is the one love above all others. As Lewis puts in the
mouth of the devil Screwtape, God "wants each man, in the long run,
to be able to recognize all creatures (even himself) as glorious and excellent things."[25] This applies not only to creatures but to our technologies. We can love Spotify and Xbox best, and even more fully, when
we love God above all. That is the proper ordering of our loves.

When our loves are properly ordered—or, more precisely, when God
by his grace helps us order our loves and our lives—we can also love
technology fittingly. So let's use technology well. But let's also be "as
shrewd as snakes and as innocent as doves" (Matthew 10:16). There I go
again, quoting Jesus out of context. But by now you know what I mean.

In the next two case studies, we look at two ways we can ask the
wrong things of science, both of which are critical for emerging adults—
sexuality and global climate change. I suspect, with the first topic at
least, I have your attention. So let's head there.

ON GLOBAL CLIMATE CHANGE AND SEXUALITY

(WHERE WE'RE TEMPTED TO ASK SCIENCE FOR THINGS IT CAN'T DELIVER)

Be praised, my Lord, with all your creatures,
Especially Sir Brother Sun,
Who brings the day, and you give light to us through him.

FRANCIS OF ASSISI

S ometimes it's good to share selective pet peeves. I hope so because I'm full of them. One of my pet peeves (if you haven't guessed it by now) is that we as Christians ask the wrong things of science. We demand things science can't deliver. We insist on absolute consensus or we blame its conclusions because there are personal lifestyle changes we need to make but resist. That's certainly the case with global climate change and sexuality.

With climate change, which is a critical issue for emerging adults, we demand that science present an absolute consensus. We see varying views on climate change as an Olympic wrestling contest between adversaries instead of a call to respond in Christian faithfulness. Instead, as believers we need to set climate change within the bigger context of Christian stewardship.

And all of this brings me to springtime in Chico, California.

Every spring, Chico State University hosts a remarkable national conference. In this Northern California town, where the almond blossoms have just signaled the ending of winter, the spring sun greets scholars, activists, students, and community leaders from our region and all over the United States who come to learn, dialogue, and become motivated about sustainability. In fact, the event is called "The Way to Sustainability." Frequently a team from Chico-based Sierra Nevada Brewing, the seventh largest brewery in the United States, with their CEO Ken Grossman, talk about the company's efforts to be one hundred percent off the electrical grid, the use of biomass in their delivery trucks, and their efficient use of energy along the way. Other speakers describe the point when our planet might hit the point of no return with greenhouse gases. Other presentations are more prosaic, like "Reducing Your Carbon Footprint Through Home Energy Conservation." I'm happy to report that Chico State is quite serious about all this—besides the sustainability conference, our university has hired a provost for sustainability, and sustainability is also one of our academic emphases, or education "pathways." The university has set an institutional goal of being carbon-neutral by 2030.

Sustainability asks the question, how do we live with our natural resources to sustain our future? One definition of sustainability is "a way of life and practices that utilize natural resources by means that can endure, thereby providing for the welfare and ecological balance of the natural world."[1] Global climate change, part of the search for sustainability, represents the scientific consensus that our planet as a whole is warming in ways that threaten a variety of life forms and that human activity is a major cause.

I have spoken at this yearly conference, once presenting a talk called "Religious Understandings of Sustainability." Gradually, however, I slipped in the old-school term *stewardship* because it represents the best way for us to conceive of how God calls us to care for the planet.

Interestingly, in venues and events like these, the church and its sacred texts are often portrayed as archenemies of sustainability. Specifically the focus is on Genesis 1:28, where Adam is given "dominion" over the earth. This "dominion mandate" has been accused of creating our environmental crisis, which of course includes climate change. Let me point to the specific, persistent challenge presented by Dr. Lynn White, professor of medieval history at UCLA, particularly his 1969 article, "The Historical Roots of our Ecologic Crisis."[2] Even the title, with its "ecologic crisis," sounds dated, but let's not be fooled: this short, pithy, fascinating piece still continues to resonate today, which is why we're considering it here. (It also reminds us that what we today call "sustainability" was once referred to as "ecology.")

To be sure, White makes many excellent arguments and presents them in a much more subtle way than many summaries would indicate. In fact, he speaks from within the church, not as an outside critic. Nonetheless, he argues that, historically—and it is important to remember that White was a medieval historian—the Christian church has been a significant player in the West in abusing our world's ecology. He contends that Christianity set itself against and ultimately destroyed the animism present in paganism, thereby making it "possible to exploit nature in a mood of indifference to the feelings of natural objects."[3] The great historian Arnold Toynbee echoes this sentiment when he writes, "The salutary respect and awe with which man had originally regarded his environment was thus dispelled by Judaic monotheism in the versions of Israelite originators and of Christians and Muslims."[4]

White particularly bases his contention on a reading of biblical texts such as Genesis 1:26-28 and its dominion mandate. White concludes that "dominion" equals domination (or exploitation) and that this element of thought in the Hebrew Bible led to the rise of modern science and technology and thus contributed to the "ecologic crisis." From the concept that "man" is "not simply part of nature; he is made in God's image," White concludes that Christianity "not only established a dualism of man and nature but also insisted that it is God's will that man exploit nature for his proper ends."[5]

White was no scholar of the Bible or of Christian doctrine, so he could have pursued a few biblical and theological traditions more deeply. First of all, the texts that speak of dominion have a much richer and more subtle meaning. "Dominion" is closely related to "stewardship," the concept that the people of Israel were to act as God's viceroy on earth, to bear "his own image," as Genesis 1:27 says. This is language of the ancient near eastern kings who set up their image to demonstrate the boundaries of their territory and how it was governed.[6] In the history of Israel, the king could be judged according to the mercy code in the Torah (the first five books of the Bible)—that is, by his concern for the least, exemplified in the phrase "widows and orphans."

In sum, stewardship is care, not domination. Humans are part of nature but also bear a distinct privilege in their power and ability to affect the natural world—I take this to be reasonably self-evident—but what are we to do with this capacity? The Bible call us to careful stewardship of creation. For this reason, some like the term "creation care."[7]

White does note the life of Francis of Assisi (1182–1226), who lived in great harmony with the natural world. Francis wrote one of the earliest poems in the Italian language, called the "Canticle of the Sun," which I cited in the epitaph and which continues,

> Question:
> How do most college students see science?
> Answer: "There's so much that's a part of science. A lot of people see the science that influences their lives. Climate change is a science topic."
>
> *TREVOR, AGE 18*

Be praised, my Lord, for Sister Moon and the Stars.
In heaven you have formed them, bright, and precious, and beautiful.[8]

Notice, with this poem, that Francis sets humankind clearly within the created world, naming the sun his "brother" and the moon and stars his "sister." In fact, the biblical traditions talk of humankind as one component of creation—even minuscule in relation to God's perspective. And yet they have a particular calling. Psalm 8's question of wonder is worth repeating:

When I consider your heavens,
    the work of your fingers,
the moon and the stars,
    which you have set in place,
what is mankind that you are mindful of them,
    human beings that you care for them? (Psalm 8:3-4)

White is probably presenting Francis as the exception that proves the rule, but that argument has to strain out a considerable amount of Jewish and Christian history, along with the thousand-year tradition of the Hebrew Scripture itself.

In Christian history, a less obvious champion of creation care is John Calvin and the Calvinist tradition generally, which has always called Christians to live simply, without ostentation, instructing its followers (to quote the Girl Scouts) "to use our resources wisely." It is the tradition in which I stand as a Presbyterian. In a broader sense, it is the tradition of which many of us, as descendants of the Puritan strain in American life, are largely unconscious heirs. In light of the current domination of market forces, it represents a stunning counter-consumerist move that— when followed—counteracts exploitation of the earth. It also offers a critique to many practices that have often characterized the history of the Christian church when we have forgotten the traditions of the

Scriptures, of Francis, of Calvin, and of many others. As Americans, we could certainly curb our consumption of fossil fuels, which is tied to global warming, by returning to simplicity. Overconsumption of natural resources is a spiritual issue. In other words, strong biblical theology and faithful Christian practice lead us to care for the earth as stewards, not to exploit it as consumers.

God's calling to be stewards of our earth is the correct context for us to address climate change. It's not news, of course, that some Americans, and a small minority of scientists, are resisting the findings that point to climate change. But why? Let's not forget that with all science, there are dissenters—that's how science works. In this case, the main question is whether the change in climate is caused primarily by human activity or a natural cycle. Nevertheless, the vast majority of scientists are convinced that climate change is occurring and that human activity is responsible. The largest scientific organization in the world, the American Association for the Advancement of Science (of which I am a member), has issued a call to take climate change seriously and urgently (see "What We Know" on the AAAS website, http://whatweknow.aaas.org). The AAAS is joined by leading Catholic voices such as Pope Francis and conservative Protestant organizations such as the National Association of Evangelicals, along with leading climate scientist and evangelical Katherine Hayhoe, who has commented, "Climate change is here and now, and not in some distant time or place. The choices we're making today will have a significant impact on our future."[9] But why does Hayhoe think climate change is caused by humans? She points out that when we look at the rise of global temperatures since 1900—in other words, the beginning of the Industrial Revolution and its release of carbon dioxide—the signs of human causes of climate change are clear.[10] It just makes sense. The realization that carbon dioxide and other greenhouse gases trap heat is not a new insight; it has roots in the nineteenth century.

Frankly, the resistance to climate change does not strike me as primarily scientific. Let's admit that one obstacle to acceptance is money. I remember my church business administrator expressing more than a modicum of resistance to the sustainability committee I was starting when he said, "The only greening we need in this congregation right now is saving money." But money alone is no reason to accept the status quo. More broadly, some resist climate change for economic reasons—and those motivated by greed need to be openly rebuked. Paul's command to his fellow Christians is to "put to death" a list of sins that ends with "greed, which is idolatry" (Colossians 3:5). On the other hand, others truly fear for their livelihood in threatened industries, such as coal, and I believe we need to be sensitive to these concerns. Some don't believe government should enforce the solution to climate change, and that particular question is not the intent of this brief case study. Instead, it leads me to close where I began: stewardship and our responsibility for the planet. Global climate change represents a pressing issue that we cannot avoid, but global stewardship involves much more. We need to concern ourselves for the poor who bear the brunt of the effects of climate change. We also need to think about the future, for our children. What earth will we leave for them? When the planet over which we are stewards is threatened by our actions, we have to reevaluate all our calculations.

As I was in the final throes of writing this manuscript, I needed a break and decided to walk to a Peet's Coffee for a midday cappuccino. As I was sipping my coffee, I overheard someone ask one of the employees, Nik, what the button on his apron—which read, "Burritos, not binaries"—meant. "Oh, it means we have to move beyond hyper-masculinity and hyper-femininity toward something more fluid." The customer, as far as I could tell, didn't want—or perhaps feel qualified—to carry the conversation further. He grabbed his drink and walked away. But I was intrigued. So I asked Nik about the button. He said,

"I'm half Native American, and in my culture, there are people who possess 'two spirits'—that is, those who identify with both male and female, who identify both with this world and with what's beyond. So for me, the button leads us to be transbinary, which is a very spiritual thing for me."

We talked a bit more before I headed back to my office, and on the way back I realized that the conversation was not only interesting to me, it signified something important. Given what I've learned about emerging adults, my conversation with Nik wasn't just about sexuality. It was about science. And that's something new. It also points to another way we ask science for things it can't deliver.

First, before we head into broader LBGTQ concerns, let's recognize that traditional positions on same-sex marriage seem questionable to emerging adults, not purely because they fail to keep up with contemporary trends (which I hear sometimes), but in light of science. In other words, science (as the argument goes) has offered definitive proof for sexual ethics. Consequently, the church's traditional positions of fidelity within heterosexual marriage and chastity in singleness are "questionable"—because they appear uninformed by the findings of science. Therefore the church is ethically suspect if not actually immoral.

However difficult these discussions are, let's not bet all our cards on science as the arbiter of ethics.

Here the problems aren't simply with people outside congregational walls. An increasing percentage of emerging adults within the church support same-sex marriage. In a Pew Research Center poll in 2001, Americans opposed same-sex marriage by a margin of 57 percent to 35 percent. Since then, support for same-sex marriage has steadily grown. Based on polling in 2017, a majority of Americans (62 percent) support same-sex marriage, compared with 32 percent who oppose it. Moreover, 74 percent of millennials—people born since 1980—are in favor of same-sex marriage.[11] Similarly, another study from 2014 found that "white evangelical Protestant Millennials are more than twice as likely to favor same- sex marriage as the oldest generation of white evangelical Protestants (43% vs. 19%)."[12]

I can only imagine that this trend will continue. (Incidentally, I'm not so happy that these pollsters focus on white evangelicals, but that's what they do and I'll work with it.) And though some denominations (such as Episcopal; Presbyterian Church, USA; and United Church of Christ) solemnize same-sex ceremonies, the majority of Christian churches today, along with a historical consensus, believe that marriage is a covenant or sacrament involving one man and one woman. The church's stance on this and related issues does set it in conflict with the mainstream culture of emerging adults. My interviews, research, and conversations have found that many emerging adults believe that scientific consensus supports the equivalence of heterosexual and homosexual marriage.

> "It's proven in science that you don't choose to be gay. Denying that makes you look ignorant."
>
> *TRACY, AGE 19*

Here, then, is my thesis, which may very well disappoint you. (Probably because I'm making a philosophical point. Philosophy, in this sense, is not entirely sexy.) Science can inform, but it cannot dictate ethics.

Indeed, ethical deliberations should not look to science as the final arbiter of truth. We should have no qualms about bringing philosophical, theological, and specifically biblical resources to bear. Put another way, integration doesn't mean domination. To integrate faith and science doesn't mean a monologue of science. Also, I'm not going to offer a resolution to our discussion of sexuality—that would take an entire book. I will say that the church is in a period of reassessment, and we cannot deny it. So, if you are involved in conversations with emerging adults, be well prepared for this topic when you address issues of science.

Let me offer a few definitions for clarity: Traditionally, sex is part of our biological constitution and has been presented as the binary options of "male" and "female." (This binarity is deeply embedded in the English language.) Gender is a more culturally conditioned category that traditionally maps onto the words *masculine* and *feminine*. For example, in ancient Roman culture, it was perfectly masculine for men to cry when they experienced loss. Nevertheless, sex and gender in my view are becoming nearly synonymous today. And that leads us to the more general category of sexuality—who we are as sexual creatures, what our sexual orientation is (whether we are intuitively attracted toward our own or the opposite sex), and how we perceive our sexual identity, which determines how we understand our identity in general. Sexual ethics is what we do with this: What is good sexual behavior? Should we engage in sexual intercourse before marriage? Do we stay sexually faithful to our spouse? Do we watch sexually permissive videos, television, or films? And so on. To state what is probably obvious, here we are particularly concerned with a Christian sexual ethic.

An additional sexual-identity designation today is LBGT, which stands for lesbian/bisexual/gay/transgender. Those who are primarily attracted to the other sex are considered heterosexual, while those attracted to the same sex are homosexual or gay for men and lesbian for women; those attracted to both sexes are bisexual. Transgender people are those who experience gender dysphoria. The American Psychiatric Association offers this definition: "Gender dysphoria involves a conflict between a person's physical or assigned gender and the gender with which he/she/they identify. People with gender dysphoria may be very uncomfortable with the gender they were assigned at birth."[13] Note that many would contest the word *assigned* here because, for the majority, biological sex is determined by physical makeup. Still, about one in every 1,500 to 2,000 babies is considered intersexual at birth, meaning these infants have no clear biological sex—they are born with a reproductive or sexual anatomy that doesn't fit the typical definitions of female or male.[14] Finally, some people add Q, for "queer" or "questioning," to the designation; thus LBGTQ. While many individuals have a hardwired sexual orientation that exists before they are able to choose, orientation is variable for those whose gender is not fixed.

Statistics for LBGT prevalence range from Alfred Kinsey's famous 10 percent for same-sex orientation to those put forth by UCLA's Williams Institute, a sexual orientation law and public policy think tank, which estimates that nine million Americans, or about 3.8 percent, identify as gay, lesbian, bisexual, or transgender.[15] From my perspective, whether it's 3.8 percent or 10 percent, it's still millions of people. And even it were far fewer, we'd still need to address this issue.

Now to science and sexual ethics: What does the science mean for our ethics?

Ted Peters wrote a key book on modern science and Christian doctrine called *Playing God?* He takes on the topic of genetics and whether a "gay gene" determines our sexual orientation.[16] If we conclude that it does, this raises huge ethical issues because it commits us to genetic determinism, the idea that our genes fully determine our identity and behavior. It's a kind of "nothing buttery" (as the scientist-theologian Arthur Peacocke quipped): we are "nothing but" our genes.[17] If our genes fully determine our behavior, then what do we do with bad behavior? If someone expresses violent resistance to the LBGT community, isn't that also simply genes at work? The problem with genetic determinism is that it can lead toward endorsing all kinds of behavior. To be clear, I'm not implying that same-sex attraction is morally equivalent to bias or violence, but this is one way philosophers work—it's called a *reductio ad absurdum* argument, where we carry out the implications of a position to its extreme to determine its merits. Genetic determinism, especially given the freedom Christ promises, is insufficient for a Christian sexual ethic. This, I believe, is bound up with

Jesus' promise that we are no longer slaves to sin: "So if the Son sets you free, you will be free indeed" (John 8:36).

Peters recently asked me to present some of my research on emerging adults to a class of master of divinity students. After the class, we chatted in a room at Pacific Lutheran Seminary—perched high in the Berkeley Hills with a magnificent view of the San Francisco Bay Area—and Peters told me he'd brought the findings of his work on genetics and determinism to a prominent Stanford law professor who fully promotes LBQT equality. She expressed the following concerns: if we affirm genetic determinism, then this leads to biological essentialism, to the view that our sexual preference is biological, natural, and essential to one's sexual identity. This traps a person into a sexual identity determined at birth. This woman, a lesbian by choice, wants to make social and legal room for individuals to decide whether they are heterosexual or homosexual. Thus she takes a stand against genetic determinism. Indeed, the concept of choice is central to American law.[18] My argument therefore does not lean toward liberal or conservative conclusions but simply points out the weakness of science as the arbiter of ethics.

Since we're on the topic of essentialism, it's vital to focus on a particular question: What is our essence as human beings? And even more particularly, what defines our sexual orientation and practice? Is it our genes? Many philosophers like to distinguish between what is essentially unchanging about us and what is nonessential or "accidental." So, to use Aristotle's rhetoric, we can fashion a chair out of wood or metal or a combination of materials, but this is "accidental" to its essence as a chair. It is still a chair regardless of the material it's made of. To put it more technically, an accident is a property that has no necessary connection to the essence of the thing being described. This has very powerful Christian significance. It is accidental that we are tall or short, American or Chinese. It is essential that we bear God's image (see Genesis 9:6; James 3:9) and that we are unconditionally loved by God: "I have loved you with an everlasting love" (Jeremiah 31:3).

Knowing how to apply this to our sexuality and sexual behavior is tricky but critical in our conversations with emerging adults. The position that is popular with eighteen- to thirty-year-olds (according to surveys and my own conversations) is what I just discussed: that our genes fully determine our sexuality and sexual behavior—in other words, biological or genetic essentialism ("The genes made me do it"). But as we've already discussed, it's a mistake to pin our behavior entirely on science. We need to understand the genetic correlations (and scientific findings more broadly) that exist in relationship to sexual expression, but I believe they are insufficient for our ethics, sexual or otherwise. I call on the best ethical and biblical minds in the church to keep engaging these questions with Scripture, theology, and sound ethical reasoning (philosophy), using science as a guide but not a dictator. Let's not capitulate all of our theological ethics to science.

Guidelines for Further Reflection

It's worth mentioning that the Christian response to sexual ethics today is varied and lively. Put thought leaders such as Rachel Held Evans, Justin Lee, William Loader, Megan K. DeFranza, Wesley Hill, Stephen R. Holmes, Jen Hatmaker, Tim Keller, Eugene Peterson, and Rob Bell all in one room, and you'd be in for quite the conversation. In fact, I'd direct you to any of their books, blogs, or videos for thoughtful dialogue about sexual ethics.[19]

Although I'm convinced that science can inform but not dictate ethics, nevertheless, let's be sure to take in what science can tell us. Let's seek integration where possible but head toward independence when necessary. For example, let's consider scientific findings about the genetics of same-sex attraction in studies of identical twins. (There is some correlation among twins, by the way, but it's not 100 percent.)

Also, physicians tell us that sexuality is not an entirely binary reality, as indicated by intersexuality.[20] These are topics worth pursuing, but I have constrained myself here. My hope is that we calibrate our expectations for what science can deliver in this area.

On the topic of our call to care for the earth in light of global climate change, I've argued for a particular strategy: Let's move away from a focus on climate change to the broader concerns of stewardship (or, if you prefer, creation care). I advocate this shift partly because the term "climate change" has become overpoliticized, with more Democrats subscribing to its reality and more Republicans expressing skepticism—a result of varying views on whether political solutions are needed.

But I want us not to lose sight of all the other ways that reducing pollution, recycling materials, and lowering our carbon footprint—"greening" our lives—are simply good Christian spiritual practices. We can learn to decrease our use of fossil fuels. We can make changes in our congregations. Many churches have adopted creation care as a part of their ministry. Many denominations, like my own Presbyterian Church USA, have a carbon-neutral statement.[21] (I also admit that statements by denominations, though important, have limited effect.)

This is the flip side of my central concern that we not expect too much of science: let's not expect science to make the changes necessary in our lives as (generally) wealthy US Christians who are wedded to consumption. That's something the transforming work of the Holy Spirit has to do. It is the hope I hear in Paul's stirring words: "And we all, who with unveiled faces contemplate the Lord's glory, are being transformed into his image with ever-increasing glory, which comes from the Lord, who is the Spirit" (2 Corinthians 3:18).

Finally, I have to confirm that as Christians we believe Jesus might return at any moment. And when he comes like "a thief in the night," according to 1 Thessalonians 5:2, we want to be found caring for a world that our children—my two daughters and their millennial colleagues—will inherit. It's not hard for me to imagine that one of Jesus' questions will be this: "How have you taken care of this planet that I entrusted to you?"

As we seek to remain faithful to Jesus Christ, we should read both the book of nature and the book of Scripture. How that moves us forward is the subject of our final chapter.

8

MOVING FORWARD

But one thing I do: Forgetting what is behind and straining toward what is ahead, I press on toward the goal to win the prize for which God has called me heavenward in Christ Jesus.

PHILIPPIANS 3:13-14

In a fit of pique, when his designers were asking for just a little more time to continue their creative process even if it delayed the release of the first Apple personal computers, CEO and founder Steve Jobs burst out (as he was wont to do) with a phrase that's become iconic: "Real artists ship."

I treasure this phrase because it emerged from the mouth of the man who rethought computers (the Mac), industrial and technological design (the iPhone, iPad), and retail sales (the Apple store). His brain was able to reconceive a stunning range of products. And yet this artist also realized that at some point we have to actually deliver the goods. We have to take steps. We have to ship.

So too for the church. It's a worthy endeavor to consider the issues of the past few chapters and the major concerns of emerging adults on Christian faith and mainstream science. But Christian leaders who are in ministry with eighteen- to thirty-year-olds need to know: What do we do now? What's our goal? And who can help us?

I start with the apostle Paul, whose theological reflection was impeccable but who always had the next steps in view. Even when his mind could have become restricted by the walls of his imprisonment and the prospect of imminent death, he looked to what lay before him. Because of Christ he found that he could strain like an athlete toward the prize bestowed by the leader of the games—even the emperor himself. He kept leaning toward the purpose of glorifying Christ, the true Emperor of emperors. That is our goal as well.

With Paul's inspiring words in mind, here we'll close with a few reflections that I believe are necessary for the church's mission as it moves forward in integrating faith with science.

Issues That Make a Difference

We have to make this topic relevant because, in my experience, emerging adults today have so much information to sift through that I've noticed a tendency toward pragmatism over theoretical speculation. We need to broaden the conversation to include, for example, the love for the natural world that Christian faith gives us, which is the basis for all scientific endeavor. Belief in the God who creates the natural world leads Christians to observe and enjoy nature.

This means that we affirm for emerging adults the value of exploring of new ideas and integration of all forms of knowledge—in this case, knowledge of faith with the insights of mainstream science. There is a natural openness to exploration that comes not only with the psychological development of emerging adulthood, but also with the cultural environment of college study and discovering a new career. Seek to collaborate with science-related events at local colleges or universities. If, for example, a scientist is presenting a lecture, this could be an opportunity for small group discussion.

Emerging adults generally want to feel that "yes, this topic makes a difference." This is true of the topics we've discussed such as technology, neuroscience and the soul, global climate change and sustainability, and the calling of Christians in the sciences.

Above all, we need to encourage our college students, postcollege emerging adults, and children and youth to study science, to take that wonder at the beauty and intricacy of creation to the next level. (Remember Ella and her tribe in the first chapter?) One of our most critical tasks is this: to help them see science as a legitimate Christian calling. It's no wonder that materialism creeps into today's science when there has often been an antiscience bias in the church.

But how do we know exactly which topics and trends to address? Ideally we'd all have degrees in science, theology, philosophy, or all three. But this is not the reality. Most of us who lead emerging adult ministries—or simply care to talk with eighteen- to thirty-year-olds— learn science on the side or as a component of our wider ministry skills. We have to be familiar with the key influencers and to know the top-ten hits on the science charts. Youth workers and theologians Andrew Root and Erik Leafblad advocate, "While it is unreasonable to become scientists, as youth workers we need to become knowledgeable about the ways in which science helps frame reality for our students."[1] Ministry leaders would do well to read the *New York Times'* "Science Times," attend public lectures at local colleges, or read "best science and nature writing" collections.

Be sure that you and your emerging adult– ministry leaders, even if they're not specialists in science and faith, communicate the importance of engaging mainstream science. It's critical to recognize the importance of science in the lives of college students and the fact that many of them are pursing degrees in science and future

> "Considering that you're a man of science and you're following Christ —that's cool."
>
> **TIMOTHY, AGE 19**

science-related work. Connect with this pressing life issue—namely, one's job. It's worth recalling that 52 percent of youth group teenagers will go into a science-related field;[2] from this we can extrapolate a similar percentage for college group attendees. If half of our college students and postcollege emerging adults will be involved in science-related fields, they need to know how to do their work while following the upward call of Christ. Again, let's teach the collaboration, not the controversy. (In fact, much of mainstream science has no real controversy.)

The research teams I've been part of have discovered that emerging adults are interested in integrating science and faith when either a qualified presenter or excellent material is part of the equation. In qualitative interviews, we've also found remarkable curiosity in how to bring these together. Put another way, atheist, materialist science has not entirely won the day, and only a small percentage of people adhere to such a position. Sociologist Jonathan Hill estimates this constituency at around one-sixth.[3] The strategy for this integration is to connect it with pressing life issues through relationships of trust, through skilled communicators, through a robust biblical hermeneutic, and through the use of high-quality and high-impact resources.[4]

It's also important to know where honest science finds its own limits (versus popularizers who use science to promote their own aims) and how it invites deeper wonder and worship. Science may just be the intro to a much grander song. Root and Leafblad note, "Science can be one piece of a broader and ongoing invitation to wonder and adventure, as well as doubt and uncertainty in life with God."[5] In many cases, wonder and awe at the natural world God has created lead to deeper worship. In others, investigation of science leads to a realization that there are limits to scientific insight and discovery; faith and wonder in the God beyond the natural world are the only reasonable responses.

Only God is God, and his book of Scripture complements but is certainly beyond the book of nature. Blaise Pascal, who lived in the seventeenth century and contributed to the birth of modern science, knew the limits of scientific and philosophical reason: "If we submit everything to reason our religion will be left with nothing mysterious or supernatural. If we offend the principles of reason our religion will be absurd and ridiculous. . . . Two excesses: to exclude reason, to admit nothing but reason."[6]

Engaging Endorsers

Lest we think this integration of mainstream science and Christian faith is all about arguments and rational discussions, let us recall the dictum, "We engage people with arguments, not arguments in abstraction." This also implies that changing minds is not an easy task. We have to realize that many of our eighteen- to thirty-year-olds hear about the conflict out there. They sense that science and faith don't play in the same sandbox. So we need to find voices that endorse the connection.

We in the church or in parachurch ministry have one key advantage. Encouraging the integration of mere Christianity and mainstream science isn't simply about knowledge but, as researchers have discovered, "intuitive cognitions" or "feelings of certainty" that lead to decisions about what is true—the acceptance of evolution, for example.[7] As surprising as it sounds to me (since I've been a pastor, and most of us in this profession wonder how much people really listen to us), a ministry leader's voice offers feelings of certainty that are central to defining a social world and thus what should be considered true or not. Here we turn back to Hill, who notes that trusted voices are critical for opening emerging adults to explore mainstream science: "For most students, then, it matters little what their professor teaches. . . . What their friends, parents, and pastor thinks is going to be far more important, because their social world is inextricably tied up with these significant others."[8] We believe things because those around us make them believable.

Our thinking is indeed bound up with our group. Jonathan Haidt's *Righteous Mind* is helpful in this area. Haidt, who describes himself as a Yale-educated, East Coast secular liberal, couldn't understand why people disagreed so vehemently in our country and wanted to find out more. (Incidentally, he has some of most sensitive analysis I've read of the "other side"—which for him consists of conservatives.) While ultimately I think Haidt leans too much on emotions, I still find his image helpful: that if we are riding an elephant, the "rider" is our reason and our emotions and intuitions are the elephant. And that beast is defined by our in-group, especially the six to twelve people who represent our "posse" or inner circle.[9] As a result we often think, and even more, argue, not principally to find the truth but to convince those around us that we're making sense.[10] Haidt is probably pushing his conclusions too far, but this analysis convinces me that our thinking—and our willingness to engage new ideas—is bound together with our closest community.

The importance of trust and our social world brings me to the need for the "endorser": a trusted voice in one's key group who affirms the need to integrate faith and science. Consider who could fill the role of endorser in your ministry groups. Identify emerging-adult ministry leaders who are excellent communicators and help make them more skilled in their work. Also, consider that a senior pastor's endorsement of science or a trusted scientist's positive comments on integrating science with Christian faith can have an extremely positive effect. Think about bringing in engaging outside speakers, highlighting resources on the web, and creating discussion groups around these events.

Sometimes we can also rely on leading national figures like Francis Collins, Elaine Howard Ecklund, Karl Giberson, and Jennifer Wiseman to speak on issues of faith and science. When someone is a stellar communicator and a superb scientist, it's powerful. (And we don't always have to invite them to speak. We can simply show one of their videos and facilitate a discussion.) But also get to know the local thought leaders. This is critical for college ministries—become familiar with scientists in local congregations and those in university positions nearby. Attend a talk at a library, community center, or local college.

Let's not forget that we are some of the most qualified people to speak in our own context. That's the reason we created Scientists in Congregations—to encourage scientists who are part of a Christian community (a congregation, a small group, a fellowship group) to talk about their science in the context of faith. If, for example, you see Robert talking about evolution and Robert was a great youth group leader for your son, you're much less likely to think he's out to undermine the Christian faith. If Ellen is a cognitive scientist who leads the lay counseling ministry and she describes the close connection between mind and brain, we're less inclined to believe she's challenging the existence of the soul. In other words, we can't learn everything about evolution from Richard Dawkins and about neuroscience from Stephen Pinker.

It's indispensable that emerging-adult ministry leaders, even if they are not specialists in science and faith, communicate the importance of engaging mainstream science. Emerging adults do not have as many ready-made structures for dialogue as other demographics, such as youth, do, so it takes effort to locate trusted voices and local thought leaders on science and faith and bring them into the discussion. Still, I'm convinced that with just a bit of effort, you'll find them.

> On watching a Veritas Forum talk by Francis Collins: "Collins gave proof that it is possible to believe in both [science and Christianity] since he's a scientist and a Christian."
>
> *MAISIE, AGE 19*

Identifying Translators

The endorser's task can include being a translator. But these roles can also be separated. To translate is no mean feat—it takes effort to grasp the meaning of two-thousand-year-old Gospels; it requires work to find the significance in Aristotle or Lucretius. This difficulty is not unique to Christian Scripture, but it's what any good biblical expositor has to do in the classroom or pulpit.

The basis of this task of translation is that God uses the church to speak in languages people can understand, the languages they dream in. Whenever I've had the opportunity to preach on Pentecost, the birth of the church profoundly moves me. God's strategy is stunning: one of most surprising elements of Acts 2 is that everyone who heard the message that day knew Greek, so God could have let Peter preach in that language. But to most it was a foreign tongue, forced on them by the oppressive Roman imperial government (and before that by the Hellenizing efforts of Alexander the Great). So instead of Greek, God's Spirit spoke a message to the people in their own native tongues.

Our native tongue is the language of our dreams; it is the tongue we use to cry with despair and pain as well as to shout with joy. "Then how is it that each of us hears them in our native language?" (Acts 2:8). As we take in the power of Pentecost, we realize we need to preach the gospel in the vernacular. What if we worked harder at presenting the gospel in the language people work and dream in? For many emerging adults, that's the language of science and technology. It's also true for those who live in a science- and technology-saturated world. Meaning all of us. That's the water we swim in.

However, we lack skilled translators who can present a sound approach to science and Scripture. I realize, of course, that C. S. Lewis wasn't a scientist, but he did grasp the effects of science on the wider culture and expertly articulated mere Christianity in that cultural context. And we could use more of his ilk. He too designated his work as "translating" and asked a question that has not been satisfactorily answered: "People praise me for being a translator. But where are the others? I wanted to start a school of translation."[11] Where indeed are the translators who understand the glories, challenges, and intricacies of science and can bring Christianity to a scientifically and technologically saturated age? Part of my work has been to identify these translators, but there's much left to do. Could that be part of your calling and passion?

I've discussed this already in chapter four, but it bears repeating here: in order to make ancient texts contemporarily relevant, a great deal depends on how we look at the Bible. Paul says to Timothy, "Do your best to present yourself to God as one approved, a worker who does not need to be ashamed and who correctly handles the word of truth" (2 Timothy 2:15). Just as Paul took up the Stoic word *self-sufficient*—often translated as "being content"—and put it into a Christ-key, reminding us that we are only sufficient when we do all things through the strengthening one (see Philippians 4:13), we too can learn to translate the Scripture using the words of science and technology today.

We should also encourage the church to connect emerging adults with mentors, particularly emerging adults in science-related fields with practicing scientists. As Christian Smith and Patricia Snell note, generally emerging adults would like to have mentors but have difficulty finding them.[12]

Cultivating Resources

For those who want to affect this conversation of faith and science for emerging adults, there are a number of excellent resources. Books are not entirely dead, but online content and video are essential. The emerging adult culture, nurtured on rich video content, needs visuality for communication. In addition, it's important to keep creating and expanding our resources. Those who work with video and web development will naturally be most likely to assist in these efforts. Accordingly, it's crucial to develop robust websites, maintain them regularly, and market them through social media channels. In surveys I was surprised to find that the large majority of respondents, when I asked, "If you wanted to find out more about religion and/or science, where would you go?" said they would look to the Internet. So let's engage the Internet, first by making content that will find traction, then by working through the resources that integrate mainstream science and mere Christianity.

I'll start with the ones I know well or have worked on: Scientists in Congregations (scientistsincongregations.org) and Science and Theology for Emerging Adult Ministries (thesteamproject.org). I also recommend BioLogos for its superb content (biologos.org; full disclosure—I'm on their advisory council). Cambridge University's Test of Faith (testoffaith.com) was a central component of our SEYA study with emerging adults.

God and Nature magazine (godandnature.asa3.org), the American Scientific Affiliation's online magazine, includes both essays and poetry as a way to promote conversation between science and theology. The American Scientific Affiliation itself (network.asa3.org) provides a safe atmosphere for Christians interested in science to network with each other and discuss mainstream science. They also have an active presence among students at universities across the nation. Christians in Science (cis.org.uk) in the United Kingdom is similar to the ASA in the United States. This mostly British organization welcomes anybody interested in the intersection of science and faith, although it caters to professionals. The Canadian Scientific and Christian Affiliation (csca.ca), also similar to the ASA, is a primarily Canadian fellowship that brings Christians interested in science together to integrate the Christian tradition with the best of science in an interdisciplinary conversation.

Finally, on the theological side of the ledger, the Plpit Initiative, also called Science in Sermons (www.plpit.com), housed at Fuller Theological Seminary, helps pastors bring science into their sermons. It summarizes scientific research, makes theological connections, and offers ideas to incorporate research findings in congregational life. The Ministry Theorem (ministrytheorem.calvinseminary.edu), developed by Calvin Seminary, pools manifold resources, including books, organizations, lectures, and websites that congregations can use. (It also has a great collection of resources for children and teenagers.)

Of course, that's just the beginning, and given the nature of the Internet, as soon as this list is made, new resource will be available shortly thereafter.

Telling Better, True, and Beautiful Stories

I close with this: if you related at all to my narratives in the first chapter, then you'll understand what I mean when I say we need to tell a true and better story about faith and science. This doesn't mean, "Let's make up a story." It means we have to engage the convincing, true narratives that that God has written in the book of Scripture and in the book of nature. In order to portray the specific human contours of eighteen- to thirty-year-olds' approach to faith and science, I close with vignettes from three emerging adults. I want to learn how to help them write a better, true story.

Beth, age twenty, is an energetic, engaging young woman who works hard at school as well as her jobs (to support her school costs because her parents aren't supporting her financially). She talks with confidence, but you can tell there's a hurt and defiance there too. Beth was raised in church and in a family very committed to the church, and she attended parochial school while growing up. There she asked her teachers a number of questions about her faith, including how the Bible and Christianity relate to the Big Bang and evolution. She was told to come to the principal's office for a meeting with her parents and was later invited to leave because she "asked too many questions." Not surprisingly, she left the school and the faith.

At this point Beth can't imagine being back in a congregation. "I don't think I'll ever go back to church," she stated bluntly but not unkindly. "And although I consider myself spiritual in some way, I don't see religion as an option." Asking questions about science and faith meant, in Beth's context, that she was unfaithful.

Jim is a really good guy. Though he's tall, muscular, and a bit imposing, there is a gentleness about him. His smile comes easy. His parents divorced when Jim was two, and his mother had the family go to church for a while—off and on, when she felt she needed to go. When she later remarried, Jim's stepfather never said anything about faith, and the family attended rarely. Nevertheless, Jim attended church youth camps. "I found it kind of weird," he told me, "that wherever we went, everyone was so sure. They never asked if you had any doubt."

Not able to discuss his questions and doubts, Jim drifted away. Today, a junior in college, he watches scientific atheists on YouTube (Dawkins, Harris, Hitchens, Bill Maher). He also watched the famous 2014 debate between creationist Ken Ham and Bill Nye, "the science guy." "Here's the weirdest thing," Jim said. "When Bill Nye asked Ken Ham, 'What would you need to change your mind?' Ham replied. 'Nothing.'"

Jim's mouth gaped open for a moment before he commented, "This is what turned me off—crazy stuff that religious people do. They're kind of brainwashed."

Jim is now a committed atheist. Yet as we talked, he brought up an idea he learned about in childhood: eternal life. I did what I was taught as an interviewer and stayed neutral, offering affirming statements like "I can understand what you're saying." But then he took the conversation in a different, deeper direction.

"You know, I long for heaven still—that sense that there's more to come," Jim said. "But I just can't believe it." At that moment I was asking questions, not offering answers. So I remained silent. Still, I felt a deep wistfulness in Jim—he wants Christian hope to be true, but he doesn't see a way to believe in the gospel without losing his mind.

Dave, now an emerging adult in his late twenties, was raised in Northern California. As he puts it, "I was blessed by excellent public schools." In first grade he had his first science class, and his interest grew throughout elementary school. In middle school the classes got more difficult while the experiments got more interesting. In high school the studies went deeper, focusing on biology, chemistry, physics, and earth-space. "All the while my faith in Jesus was developing, my heart yearning to love him and be his," Dave said. "As both my faith and science studies merged, however, I grew increasingly silent. By high school I was unsure how the two worked together, or if they worked together at all."

Dave continued, "I'll never forget the day I spoke up in science class. We had a new teacher, 100 percent Italian with a thick accent and brilliant mind. She referred to the Big Bang without the word 'theory.' One day I politely raised my hand and asked, 'Isn't the Big Bang still in fact a theory [by which he meant an untested conjecture]?' You could hear a pin drop it was so silent. From that moment on she referred to the Big Bang as a theory; however, and I began to naively discard science in the name of faith."

Months later at a worship concert with Matt Redman, God redirected Dave's mindset. Midway through the concert, speaker, author, and pastor Louie Giglio took the stage. His theme was "How Great Is Our God." He shared Scriptures and pictures of the universe and communicated how scientific insights revealed how grand, wonderful, and big God really is.

"Giglio told us that, while the moon and the sun appear the same size, the sun is actually four hundred times larger," Dave said. "Yet it appears the same as the moon because it is also four hundred times farther away.

If Earth were a golf ball, the sun would be fifteen feet wide. He also described a star named Betelgeuse, which is 427 light years away and twice the size of the earth's orbit around the sun. If the Earth were a golf ball, we could fill the super dome three thousand times!"

Dave later commented, "It was one of those perfect nights where I fell in love with God all over again. Even more important, I no longer viewed the insights of science as an enemy or a conflict for my faith. For the first time, the insights of modern science helped me know my God better."

That's why I believe bridging the divide between mere science and Christian faith is a vital task. It's for Jim, for Beth, and for Dave that I want us to write better, true stories. To be honest, I don't know how to write this story by myself. It's something we as the church need to do together.

I do know, however, that these true, better stories are also beautiful. They will bring together the goodness and truth of the good news with the beauty of God. There truth becomes beautiful. And it should not be overlooked that rhetoric—as an engagement with beauty—should be used in concert with philosophy—as the pursuit of truth. Truth is only worth engaging if it's beautiful, and beauty is that which allures us.

This is a particular beauty, the beauty of life making sense, of satisfying our need for deep, abiding happiness, for Aristotle's "human flourishing" and for Jesus' promise of abundant life (see John 10:10). These ancient, wise voices indicate that this obviously is not a new idea, and here I concur with the great twentieth-century French physicist Henri Poincaré:

> The scientist does not study nature because it is useful to do so. He studies it because he takes pleasure in it; and he takes pleasure in it because it is beautiful. If nature were not beautiful, it would not be worth knowing and life would not be worth living.[13]

Let's can join hands with Poincaré and with the ancient theology's love of beauty, or *philokalia*.

Our final goal is this: to weave together mainstream science and the good news of mere Christianity into a narrative that's truly beautiful and beautifully true.

ACKNOWLEDGMENTS

I would like to thank all of those who participated in two projects designed to engage mainstream science and mere Christianity (often with emerging adults): Scientists in Congregations (or SinC) and Science and Theology for Emerging Adult Ministries (or STEAM). My gratitude extends as well to Bidwell Presbyterian Church and the dream team at Fuller Theological Seminary's Office for Science, Theology, and Religion Initiatives (STAR) for housing these projects, as well as to the John Templeton Foundation for providing the financial support for this work. Finally, to my project coleaders, Dave Navarra and David Wood, as well as Drew Rick-Miller, who has worked on "both sides of the aisle." Thank you!

FURTHER READING

I realize that the campus worker or pastor for twenty-somethings isn't saying, "Wow! I've got so much extra time on my hands! I wonder if there are a hundred or so books to read on these topics." Instead of every book I've cited here, I'll offer my top-ten list.

General Books on Science and Faith

Collins, Francis. *The Language of God*. New York: Free Press, 2007. Even after a decade, this is still one of the finest books from one of the finest minds in science on relating mere Christianity and mainstream science. It focuses on genetics, which is appropriate since that's one of Collins' areas of specialization.

Lewis, C. S. *Mere Christianity*. New York: MacMillan, 1952. Although not a book about science and faith per se, Lewis takes up scientifically informed atheism and responds with a beautiful articulation of faith. Even after decades, he's still frequently quoted. I'm going to cheat here and add, as a subpoint, my book *C. S. Lewis and the Crisis of a Christian* (Louisville: Westminster John Knox, 2014). Blame one of my reviewers for recommending that I include it.

McGrath, Alister. *Religion and Science: A New Introduction*. New York: Oxford, 2010. This is a more comprehensive overview of various topics and figures in the field. McGrath has all kinds of degrees, two of which are in science and theology, and his range of scholarship is impressive.

Books on Emerging Adult Culture

Hill, Jonathan. *Emerging Adulthood and Faith*. Grand Rapids: Calvin College Press, 2015. It's brief but packed with relevant insight and a bit less alarmist than many.

Kinnaman, David, with Aly Hawkins. *You Lost Me: Why Young Christians Are Leaving Church . . . and Rethinking Faith*. Grand Rapids: Baker Books, 2011. This book set the agenda for many of my reflections here. It raises alarm bells on what the church needs to do in light of emerging trends. Kinnaman is president of the Barna Group, which also maintains a website with current trends, www.barna.com.

Smith, Christian, with Patricia Snell. *Souls in Transition: The Religious and Spiritual Lives of Emerging Adults*. New York: Oxford University Press, 2009. This has become a classic in the field.

Wuthnow, Robert. *After the Baby Boomers: How Twenty- and Thirty-Somethings Are Shaping the Future of American Religion*. Princeton, NJ: Princeton University Press, 2007. Wuthnow is a premier sociologist of religion, and his is my favorite rendering of emerging adult culture and attitudes.

Books on the Bible and Evolutionary Science

Barrett, Matthew, and Ardel B. Canedy, eds. *Four Views on the Historical Adam*. Grand Rapids: Zondervan, 2013. This does just what the title says and helps us see how to relate modern evolutionary science and the historical Adam (and Eve).

Fugle, Gary. *Laying Down Arms to Heal the Creation-Evolution Divide*. Eugene, OR: Wipf & Stock, 2015. This book, from a master teacher, covers the landscape of why evolutionary thought makes sense and what it means for biblical interpretation and theology.

Walton, John. *The Lost World of Adam and Eve: Genesis 2–3 and the Human Origins Debate*. Downers Grove, IL: IVP Academic, 2015. This is just one of many resources that Walton, as an Old Testament scholar, has written and is the most relevant to the concerns of this book.

To state the painfully obvious, there are many, many more. But I limited myself to ten (more or less). If you want to go further with other topics I've discussed, just follow my notes for additional titles that are worth your time.

NOTES

1 Creation, Beauty, and Science

[1]This I learned from David Bentley Hart in *The Beauty of the Infinite* (Grand Rapids: Eerdmans, 2004), 30.

[2]Jonathan Edwards, *A Jonathan Edwards Reader*, ed. John E. Smith, Harry S. Stout, and Kenneth P. Minkem (New Haven, CT: Yale University Press, 2003), 252.

[3]I am adapting a phrase Andy Crouch used to title an article of mine: Greg Cootsona, "When Science Comes to Church," *Christianity Today*, March 5, 2014, www.christianitytoday.com/ct/2014/march-web-only/when-science-comes-to-church.html.

[4]Justin Barrett, *Born Believers: The Science of Children's Religious Belief* (New York: Free Press, 2012).

[5]Augustine, *The Literal Meaning of Genesis*, ed. John Hammond Taylor, Ancient Christian Writers (New York: Hammond Press, 1982), 1:42-43.

[6]C. S. Lewis, "Christian Apologetics," in *God in the Dock*, ed. Walter Hooper (Grand Rapids: Eerdmans, 1970), 93.

[7]C. S. Lewis, "Modern Man and Categories of Thought," in *Present Concerns: A Compelling Collection of Timely, Journalistic Essays* (London: Fount Paperbacks, 1986), 63.

[8]Francis Bacon; cited in Gayle Woloschak, "The Broad Science-Religion Dialogue: Maximus, Augustine, and Others," in *Science and the Eastern Orthodox Church*, ed. Daniel Buxhoeveden and Gayle Woloschak (Burlington, VT: Ashgate, 2011), 133.

[9]David Kinnaman with Aly Hawkins, *You Lost Me: Why Young Christians Are Leaving Church . . . and Rethinking Faith* (Grand Rapids: Baker Books, 2011), 140.

[10]See "America's Changing Religious Landscape," Pew Research Center, May 12, 2015, www.pewforum.org/2015/05/12/americas-changing-religious-landscape.

[11]Kinnaman, *You Lost Me*, chap. 6.

[12]See "Church Attendance Trends Around the Country," Barna, May 26, 2017, www.barna.com/research/church-attendance-trends-around-country.

[13]Cathy Lynn Grossman, "70 Percent Of Evangelicals Believe Religion and Science Are Not in Conflict," Huffpost, March 20, 2015, www.huffingtonpost.com/2015/03/16/evangelicals-religion-science_n_6880356.html.

[14]Francis Collins, "Why I'm a Man of Science—and Faith," *National Geographic*, March 19, 2015, news.nationalgeographic.com/2015/03/150319-three-questions -francis-collins-nih-science.

[15]Quoted in Robert John Russell, William R. Stoeger, and George V. Coyne, eds., *Physics, Philosophy, and Theology* (Notre Dame, IN: University of Notre Dame Press, 1988), 13.

[16]See Thomas Oden, *The Living God, Systematic Theology*, vol. 1 (New York: HarperCollins, 1987).

[17]To read these creeds, go to http://anglicansonline.org/basics/nicene.html and www.theopedia.com/chalcedonian-creed. Keep in mind that the phrase "from the Son" was added after the original formulation. For more, see Timothy Ware, *The Orthodox Church: An Introduction to Eastern Christianity*, 3rd ed. (London: Penguin, 2015), esp. 222-23.

[18]Alsadair MacIntyre, *After Virtue: A Study in Moral Theory*, 2nd ed. (Notre Dame, IN: University of Notre Dame Press, 1984), 260.

[19]*Merriam-Webster Dictionary*, s.v. "technology," definition for English Language Learners, www.merriam-webster.com/dictionary/technology.

2 Emerging Adult Faith: Not an LP, but a Digital Download

[1]Victor Lowe, *Alfred North Whitehead: The Man and His Work, 1910-1947* (Baltimore: Johns Hopkins University Press, 1990), 2:135.

[2]Alfred North Whitehead, *Science and the Modern World* (New York: Free Press, 1997), 181-82.

[3]Ecklund, a professor at Rice University, is perhaps the world's leading sociologist in science and religion. Wuthnow, a former student of the great Berkeley sociologist Robert Bellah (go Bears!), has done my favorite study of this demographic. (Note that since Wuthnow wrote his study in 2007, some of his subjects are now older than thirty.) Smith, a sociologist at Notre Dame University, has conducted two landmark studies on emerging adults and faith. And Calvin College sociologist Hill has done some of the most intensive work on eighteen- to thirty-year-olds' experience with religion and science. These four researchers are "monsters," as we say in jazz, who in my view define the field. See Robert Wuthnow, *After the Baby Boomers: How Twenty- and Thirty-Somethings Are Shaping American Religion* (Princeton, NJ: Princeton University Press, 2007); Elaine Howard Ecklund, *Science vs. Religion: What Scientists Really Think* (Oxford: Oxford University Press,

2010); Elaine Howard Ecklund and Jerry Z. Park, "Conflict Between Religion and Science Among Academic Scientists?," *Journal of the Scientific Study of Religion* 48 (2009): 276-92; Christian Smith with Patricia Snell, *Souls in Transition: The Religious and Spiritual Lives of Emerging Adults* (New York, Oxford University Press, 2009); Jonathan Hill, "National Study of Religion & Human Origins," 2014, available at http://biologos.org/uploads/projects/nsrho-report.pdf; Hill, *Emerging Adulthood and Faith* (Grand Rapids: Calvin College Press, 2015).

[4]For some brief information on this project, which was made possible through the support of a grant from the John Templeton Foundation (JTF), see www.scientists incongregations.org/seya. The opinions in this book, by the way, are mine, not necessarily those of the JTF.

[5]Examples of these resources include Ruth Bancewicz, ed., *Test of Faith: Spiritual Journeys with Scientists* (Milton Keynes, UK: Paternoster, 2009), and Alister McGrath, *Science and Religion: A New Introduction* (Oxford, UK: Wiley-Blackwell, 2010).

[6]I invited these emerging adults to be interviewed by me largely based on a preexisting relationship; they mostly represented a separate population from the survey.

[7]Jeffrey Arnett, *Adolescence and Emerging Adulthood: A Cultural Approach* (Upper Saddle River, NJ: Prentice Hall, 2000), 469-80.

[8]David Setran and Chris Kiesling, *Spiritual Formation in Emerging Adulthood: A Practical Theology for College and Young Adult Ministry* (Grand Rapids: Baker, 2013), 2.

[9]Ibid.

[10]Wuthnow, *After the Baby Boomers*, 11.

[11]Summarized by Setran and Kiesling, *Spiritual Formation*, 3-4.

[12]Ibid., 242n12.

[13]Ibid., 165.

[14]Jeffrey Arnett, "Emerging Adulthood: A Theory of Development from the Late Teens Through the Twenties," *American Psychologist* 55, no. 5 (May 2000): 469.

[15]Setran and Kiesling, *Spiritual Formation*, 4.

[16]Alexandra Robbins and Abby Wilner, *Quarterlife Crisis: The Unique Challenges of Life in Your Twenties* (New York: Jeremy P. Tarcher/Putnam, 2001).

[17]Meg Jay, *The Defining Decade: Why Your Twenties Matter—and How to Make the Most of Them Now* (New York: Twelve, 2013).

[18]Setran and Kiesling, *Spiritual Formation*, 4.

[19]Smith, *Souls in Transition*, 6.

[20]Christian Smith, *Lost in Transition: The Dark Side of Emerging Adulthood* (New York: Oxford University Press, 2011).

[21]Wuthnow, *After the Baby Boomers*.

[22]Smith, *Souls in Transition*, 281.

[23]See "Total Number of Websites," Internet Live Stats, accessed August 7, 2017, www.internetlivestats.com/total-number-of-websites.

[24]The theologian Thomas Torrance, who worked a great deal with the natural sciences, was especially insistent on this point. See, for example, his careful articulation in *Theological Science* (London: Oxford, 1969), esp. chapter 5, "The Problems of Logic."

Case Study: Addressing the New Atheism

[1]Amy McCaig, "Most British Scientists Cited in Study Feel Richard Dawkins' Work Misrepresents Science," Rice University News and Media, October 31, 2016, news.rice.edu/2016/10/31/most-british-scientists-cited-in-study-feel -richard-dawkins-work-misrepresents-science-2.

[2]Christopher Hitchens, *God Is Not Great: How Religion Poisons Everything* (New York: Twelve Books, 2007), 150.

[3]Richard Dawkins, *The Selfish Gene*, 2nd ed. (Oxford: Oxford University Press, 1989), 330.

[4]Alister McGrath, *Science and Religion: A New Introduction* (Hoboken, NJ: Wiley-Blackwell, 2009), 149.

[5]C. S. Lewis, *Mere Christianity* (New York, MacMillan, 1952).

[6]Richard Dawkins, interview with Howard Conder on Revelation TV, March 2011, www.youtube.com/watch?v=h8PNvplrmHI.

[7]"Religion Among the Millennials," Pew Research Center, February 17, 2010, www .pewforum.org/2010/02/17/religion-among-the-millennials.

[8]See, for example, Richard Dawkins, "Militant Atheism," TED, February 2002, www.ted.com/talks/richard_dawkins_on_militant_atheism.

[9]Charles Townes, "Logic and Uncertainties in Science and Religion," in *Science and Theology: The New Consonance*, ed. Ted Peters (Boulder, CO: Westview, 1998), 46.

[10]Alfred North Whitehead, *Science and the Modern World* (New York: Free Press, 1997), 4.

[11]Richard Dawkins, *River out of Eden: A Darwinian View of Life* (New York: Basic Books, 1996), 132-33.

[12]Cited in John Haught, "Evolution, Tragedy, and Hope," in *Science and Theology: The New Consonance*, ed. Ted Peters (Boulder, CO: Westview, 1998), 229.

[13]Stephen Weinberg, *The First Three Minutes: A Modern View of the Origin of the Universe*, 2nd ed. (New York: Basic Books, 1993), 154.

[14]Ibid., 155.

[15]Cited in Nancy K. Frankenberry, ed., *The Faith of Scientists in Their Own Words* (Princeton, NJ: Princeton University Press, 2008), 336.

3 Emerging Adults: Are They None and Done?

[1]Dan Barker, *The Good Atheist: Living a Purpose-Filled Life Without God* (Berkeley, CA: Ulysses Press, 2011), 13.

[2]"DeBaptismal Certificate," Freedom from Religion Foundation, accessed August 11, 2017, ffrf.org/publications/debaptism-certificate.

[3]Joshua Packard, "Meet the Dones," *Christianity Today*, Summer 2015, www .christianitytoday.com/pastors/2015/summer-2015/meet-dones.html.

[4]"Americans Divided on the Importance of Church," Barna Group, March 24, 2014, www.barna.com/research/americans-divided-on-the-importance-of-church.

[5]Ian Barbour, *When Science Meets Religion: Enemies, Strangers, or Partners?* (New York: HarperCollins, 2000).

[6]See Geoffrey Cantor and Chris Kenny, "Barbour's Fourfold Way: Problems with His Taxonomy of Science-Religion Relationships," *Zygon* 36, no. 4 (2001): 765-81.

[7]Melissa Skinner, "Funding Darwin in the Church," *Answers Magazine*, May 1, 2017, https://answersingenesis.org/christianity/church/funding-darwin-church.

[8]Stephen Weinberg, "A Designer Universe?," *New York Review of Books*, October 21, 1999, www.nybooks.com/articles/1999/10/21/a-designer-universe.

[9]Andrew Dickson White, *A History of the Warfare of Science with Theology in Christendom* (New York, 1896); John William Draper, *History of the Conflict Between Religion and Science* (New York, 1874).

[10]Critique of the warfare thesis can be found, for example, in John Hedley Brooke, *Science and Religion: Some Historical Perspectives*, Cambridge History of Science (Cambridge: Cambridge University Press, 1991); Thomas Dixon, Geoffrey Cantor, and Stephen Pumfrey, eds., *Science and Religion: New Historical Perspectives* (Cambridge: Cambridge University Press, 2010); Gary B. Ferngren, ed., *Science and Religion: A Historical Introduction* (Baltimore and London: Johns Hopkins University Press, 2002); David C. Lindberg and Ronald L. Numbers, eds., *Where Science and Christianity Meet* (Chicago: University of Chicago Press, 2003);

Ronald Numbers, ed., *Galileo Goes to Jail and Other Myths About Science and Religion* (Harvard: Harvard University Press, 2010). One work that promotes the thesis is Roland Bénabou, Davide Ticchi, and Andrea Vindigni, "Forbidden Fruits: The Political Economy of Science, Religion, and Growth," 2013, available at www .princeton.edu/~rbenabou/papers/Religion%20December%201g_snd.pdf.

[11] Jerry Coyne, *Faith Versus Fact: Why Science and Religion Are Incompatible* (New York: Penguin Books, 2016); Richard Dawkins, *The God Delusion* (New York: Mariner Books, 2008).

[12] "There can be no scientific problems, objections or aids in relation to what Holy Scripture and the Christian Church understand by the divine work of creation." Karl Barth, *Church Dogmatics III: The Doctrine of Creation, Part One*, trans. J. W. Edwards, O. Bussey, and Harold Knight (Edinburgh: T & T Clark, 1958 [1945]), ix.

[13] "Compatibility of Science and Religion," The National Academies of Sciences, Engineering, and Medicine, Evolution Resources, 2017, www.nas.edu/evolution /Compatibility.html.

[14] Christian Smith with Patricia Snell, *Souls in Transition: The Religious and Spiritual Lives of Emerging Adults* (Oxford: Oxford University Press, 2009), chap. 10.

[15] Stephen Jay Gould, *Rock of Ages: Science and Religion in the Fullness of Life* (New York: Ballantine Books, 2002), 63.

[16] Stephen J. Gould, "Impeaching a Self-Appointed Judge," *Scientific American* 267, no. 1 (1992): 119.

[17] For a philosophical approach to this question, see Ric Machuga, *Three Theological Mistakes: How to Correct Enlightenment Assumptions About God, Miracles, and Free Will* (Eugene, OR: Cascade Books, 2015), esp. 93-99, and "The Hows of Science and the Whys of Philosophy: Why Final Causes Are Still Necessary," in *In Defense of the Soul: What It Means to Be Human* (Grand Rapids: Brazos, 2002), 57-63.

[18] Kyle C. Longest and Christian Smith, "Conflicting or Compatible: Beliefs About Religion and Science Among Emerging Adults in the United States," *Sociological Forum* 26, no. 4 (2011): 846-69, esp. 854.

[19] Christopher P. Scheitle, "U.S. College Students' Perception of Religion and Science: Conflict, Collaboration, or Independence? A Research Note," *Journal for the Scientific Study of Religion* 50, no. 1 (2011): 175.

[20] Ibid., 178-79.

[21] Elaine Howard Ecklund and Jerry Z. Park, "Conflict Between Religion and Science Among Academic Scientists?," *Journal for the Scientific Study of Religion* 48, no. 2 (2009): 276.

[22]Cary Funk and Becka A. Alper, "Perception of Conflict Between Science and Religion," Pew Research Center Internet & Technology, October 22, 2015, www .pewinternet.org/2015/10/22/perception-of-conflict-between-science-and-religion.

[23]Scheitle, "U.S. College Students' Perception," 179-80.

[24]Blaise Pascal, *Pensées*, trans. A. J. Krailsheimer (New York: Penguin, 1966), 430.

[25]Becky Ham, "Religious and Scientific Communities May Be Less Combative Than Commonly Portrayed," American Association for the Advancement of Science, February 17, 2014, www.aaas.org/news/religious-and-scientific -communities-may-be-less-combative-commonly-portrayed.

[26]For more on the findings of the SEYA study, see www.scientistsincongregations .org/seya. The results support the conclusion that the attitudes of emerging adults move away from conflict toward integration on Barbour's typology when these individuals encounter high-quality material that demonstrates ways in which religion and science can be integrated.

[27]N. T. Wright, *John for Everyone: Part One, Chapters 1-10* (Louisville, KY: Westminster John Knox, 2002), 4.

[28]Creation Education Center Facebook page, www.facebook.com/creation education/?hc_location=ufi.

Case Study: Cognitive Science and Reasons Not to Believe

[1]For a brief introduction to Pinker's thought, see the interview with him and Rebecca Goldstein in Steve Paulson, *Atoms and Eden: Conversations on Science and Religion* (Oxford: Oxford University Press, 2010), 229-44.

[2]For one excellent discussion of the biblical witness with extensive interaction with the relevant sciences, see Joel B. Green, *Body, Soul, and Human Life: The Nature of Humanity in the Bible*, Studies in Theological Interpretation (Colorado Springs: Paternoster, 2008).

[3]N. T. Wright, *The Day the Revolution Began: Reconsidering the Meaning of Jesus's Crucifixion* (San Francisco: HarperOne, 2016), 147.

[4]Justin L. Barrett and Emily Reed Burdett, "The Cognitive Science of Religion," *The Psychologist* 24 (April 2011): 252-55, thepsychologist.bps.org.uk/volume-24 /edition-4/cognitive-science-religion. Barrett's *Cognitive Science, Religion, and Theology: From Human Minds to Divine Minds*, Templeton Science and Religion Series (West Conshohocken, PA: Templeton Press, 2011), is an excellent introduction to the field.

[5]Barrett, *Cognitive Science*, 59.

[6]Ibid., 71. This feature of early childhood has been termed "promiscuous teleology" by the psychologist Deborah Kelemen (ibid., 70).

[7]Ibid., 86.

[8]Andrew Newberg and Eugene D'Aquili, *Why God Won't Go Away: Brain Science and the Biology of Belief* (New York: Ballantine, 2002). See also Andrew Newberg, *Principles of Neurotheology*, Ashgate Religion and Science Series (Burlington, VT: Ashgate, 2010).

[9]See Pascal Boyer, *Religion Explained: The Evolutionary Origins of Religious Thought* (New York: Basic Books, 2002).

[10]I have more to say on this in "Science and the *Sensus Divinitatis*: The Promise and Problem of the Natural Knowledge of God," in *Connecting Faith and Science: Philosophical and Theological Inquiries*, ed. Mathew Hill and Curtis Holtzen (Claremont, CA: Claremont School of Theology Press, forthcoming).

4 On a Crash Course with Hermeneutics

[1]See Ronald L. Numbers, ed., *Galileo Goes to Jail and Other Myths About Science and Religion* (Cambridge, MA: Harvard University Press, 2010), and, for a brief summary of the key points, Greg Cootsona, *Creation and Last Things* (Louisville: Geneva Press), 24-27.

[2]See Galileo's "Letter to the Grand Duchess Christina of Tuscany, 1615," Modern History Sourcebook: Galileo Galilei, scitech.au.dk/fileadmin/site_files/science .au.dk/NF/Komm/DenbevaegedeJord/Letter_to_the_Grand_Duchess _Christina_of_Tuscany.1615__Gallilei.pdf.

[3]John Calvin, *Institutes of the Christian Religion*, ed. John T. McNeil, trans. Ford Lewis Battles, The Library of Christian Classics (Philadelphia: Westminster, 1960 [1559]), 2.2.15.

[4]Lee Rainie and Cary Funk, "An Elaboration of AAAS Scientists' Views," Pew Research Center Internet & Technology, July 23, 2015, www.pewinternet .org/2015/07/23/an-elaboration-of-aaas-scientists-views.

[5]"Statement of Faith and Educational Purpose," Wheaton College, October 17, 1992, www.wheaton.edu/about-wheaton/statement-of-faith-and-educational-purpose.

[6]"The Chicago Statement on Biblical Inerrancy," International Council on Biblical Inerrancy, accessed August 15, 2017, library.dts.edu/Pages/TL/Special /ICBI_1.pdf. This document, incidentally, makes "infallible" (does not fail) almost coterminous with "inerrant" (does not err), which is a distinction I am drawing. Wheaton uses only "inerrancy."

[7]"Statement of Faith," Fuller Theological Seminary, accessed August 15, 2017, fuller.edu/about/mission-and-values/statement-of-faith.

[8]C. S. Lewis, *Reflections on the Psalms* (New York: Harcourt Brace Jovanovich, 1958), 109.

[9]Even though leading biblical scholar John Walton insists that it is seven days of creation because the seventh day of creation is the consummation, where God the Lord is in the temple and the temple is the world, I still defend the assertion that God made the heavens in six days. See Walton, *The Lost World of Genesis One* (Downers Grove, IL: IVP Academic, 2009).

[10]Alister McGrath calls Job 38:1-42 "unquestionably the most comprehensive understanding of God as creator to be found in the Old Testament." *Science and Religion: A New Introduction*, 2nd ed. (Malden, MA: Wiley Blackwell, 2010), 86. I must admit that this strikes me as a *slight* overstatement.

[11]This section mirrors Cootsona, *Creation and Last Things*, 4.

[12]This emphasis on order as a key component of creation and proper functioning as similarly central to the meaning of *tov* or "good" in the Hebrew of Genesis 1-3, is an allusion to John Walton, *The Lost World of Adam and Eve* (Downers Grove, IL: IVP Academic, 2015), esp. chaps. 3, 5.

[13]George T. Thompson's Facebook page, www.facebook.com/akuyperian?fref=ufi &rc=p, responding to the book edited by Kathryn Applegate and J. B. Stump, *How I Changed My Mind About Evolution* (Downers Grove, IL: IVP Academic, 2016).

[14]"How Are the Ages of the Earth and Universe Calculated?," BioLogos, accessed August 15, 2017, biologos.org/common-questions/scientific-evidence/ages-of -the-earth-and-universe.

[15]See John Walton, *The Lost World of Genesis One: Ancient Cosmology and the Origins Debate* (Downers Grove, IL: IVP Academic, 2009), 9.

Two Case Studies: Making Too Much of a Good Thing

[1]Cited in Timothy Ferris, *Coming of Age in the Milky Way* (New York: Harper-Collins, 1988), 211.

[2]Fred Hoyle, *Facts and Dogmas in Cosmology and Elsewhere* (Cambridge: Cambridge University Press, 1982), 2.

[3]Robert Jastrow, *God and the Astronomers* (New York: W. W. Norton, 1992), 125.

[4]Alister McGrath, *Science and Religion: A New Introduction*, 2nd ed. (Malden, MA: Wiley Blackwell, 2010), 152.

[5]Lawrence Krauss, *A Universe from Nothing: Why There Is Something Rather Than Nothing* (New York: Atria Books, 2013).

[6]See R. J. Russell, "Finite Creation Without a Beginning," in Robert John Russell, Nancey Murphy, and C. J. Isham, eds., *Quantum Cosmology and the Laws of Nature* (Vatican City State and Berkeley, CA: Vatican Observatory Publications and the Center for Theology and the Natural Sciences, 1993).

[7]"Fine-Tuned Universe," Wikipedia, August 11, 2017, en.wikipedia.org/wiki /Fine-tuned_Universe.

[8]Note John Polkinghorne's concerns about the term "anthropic principle" in *Science and Religion in Quest of Truth* (New Haven, CT: Yale University Press, 2011), 54-55.

[9]Freeman Dyson, *Disturbing the Universe* (New York: Harper & Row, 1979), 256.

[10]McGrath, *Science and Religion*, 155.

5 *Adam, Eve, and History*

[1]Pew does consistently solid and trustworthy work on all forms of public opinion. "Public's Views on Human Evolution," Pew Research Center Religion & Public Life, December 30, 2013, www.pewforum.org/2013/12/30/publics-views-on -human-evolution.

[2]See Matthew Barrett and Ardel B. Canedy, eds., *Four Views on the Historical Adam* (Grand Rapids: Zondervan, 2013); Greg Cootsona, *Creation and Last Things: At the Intersection of Theology and Science* (Louisville: Geneva, 2002), esp. chap. 4; Gary Fugle, *Laying Down Arms to Heal the Creation-Evolution Divide* (Eugene, OR: Wipf & Stock, 2015), esp. chap. 18; David Venema and Scot McKnight, *Adam and the Genome: Reading Scripture After Genetic Science* (Grand Rapids: Brazos, 2017); Gregg Davidson, "Genetics, the Nephilim, and the Historicity of Adam," *Perspectives on Science and Christian Faith* 67, no. 1 (2015): 25-34, www.asa3.org/ASA/PSCF/2015/PSCF3-15Davidson.pdf; C. S. Lewis, *The Problem of Pain* (New York: MacMillan, 1962).

[3]Lewis, *Problem of Pain*, 77.

[4]By the way, the preferred term is now *hominin*, not *hominid*. The hominid group consists of all modern and extinct great apes—that is, modern humans, chimpanzees, gorillas, and orangutans, plus their immediate ancestors. The hominin group consists of modern humans, extinct human species, and our immediate ancestors, including *Homo, Australopithecus, Paranthropus,* and *Ardipithecus* species. See Beth Blaxland, "Hominid and Hominin—What's the Difference?,"

Australian Museum, February 5, 2016, australianmuseum.net.au/hominid-and -hominin-whats-the-difference.

[5] Lewis, *Problem of Pain*, 79-80.

[6] Scot McKnight, speaking at the 2017 BioLogos conference in Houston; quoted in Bob Allen, "How the Human Genome Project Changed This Professor's Reading of Genesis," Baptist News Global, May 3, 2017, https://baptistnews.com /article/professor-says-human-genome-project-changed-reading-gen-1-3/.

[7] For the more detailed discussion, see Dennis R. Venema and Scot McKnight, *Adam and the Genome: Reading Scripture After Genetic Science* (Grand Rapids: Brazos, 2017), chap. 5. I recognize there are other ways to read the genomic and genea-logical record, e.g., S. Joshua Swamidass, "The Overlooked Science of Genealogical Ancestry," *Perspectives on Science and Christian Faith* 70, no. 1 (2018).

[8] John Walton, *The Lost World of Adam and Eve* (Downers Grove, IL: IVP Academic, 2015), 74.

[9] Ibid., 179.

[10] Swamidass, "Overlooked Science."

[11] Derek Kidner, *Genesis: An Introduction and Commentary*, Tyndale Old Testament Commentaries (Downers Grove, IL: InterVarsity Press, 1982), 31.

[12] Ibid., 32.

[13] Gary Fugle, "Where Is God in Nature?," BioLogos, October 29, 2015, biologos .org/blogs/ted-davis-reading-the-book-of-nature/where-is-god-in-nature.

[14] Fugle, *Laying Down Arms,* 252.

[15] Ibid., 253.

[16] For what it's worth, this phrase, which is literally "because of one, all sinned"—*eph hō pantes hēmarton*—is the crux of the issue historically. See C. E. B. Cranfield's careful exegesis in *Romans*, International Critical Commentary, ed. J. A. Emerton and C. E. B. Cranfield (Edinburgh: T&T Clark, 1975), 274-81.

[17] "Statement of Faith and Educational Purpose," Wheaton College, October 17, 1992, www.wheaton.edu/about-wheaton/statement-of-faith-and-educational-purpose.

[18] Just in case it seems that I'm skating over N. T. Wright and others' interpretation that Paul is describing his pre-Christian experience, I'm aware of that school of thought but am convinced by other commentaries, ably represented by Cranfield, *Romans*.

[19] James D. G. Dunn, *Romans 1-8*, Word Biblical Commentary 38A (Dallas: Word Books, 1988), 289.

[20] Ibid.

[21] Ibid., 90.

[22] Tim Keller, quoted in Richard N. Ostling, "The Search for the Historical Adam," *Christianity Today*, June 3, 2011, www.christianitytoday.com/ct/2011/june /historicaladam.html.

[23] There are twenty-two places where Adam is a name (not a town as in Joshua 3:16 or Hosea 6:7), but principally this occurs in Genesis 2–3, Romans 5, and 1 Corinthians 15. Here's the full list: Genesis 2:4; 2:20, 25; 3:17, 20, 21; 4:1, 25; 5:1, 3, 4, 5; 1 Chronicles 1:1; Luke 3:38; Romans 5:12, 14; 1 Corinthians 15:22, 45; 1 Timothy 2:13-14; Jude 1:14.

[24] See N. T. Wright's "Excursus on Paul's Use of Adam," in Walton, *Lost World of Adam and Eve*, 172.

[25] Henri Rondet, *Original Sin: The Patristic and Theological Tradition*, trans. Cajetan Finegan (Shannon, Ireland: Ecclesia, 1972), 25.

[26] Tatha Wiley, *Original Sin: Origin, Developments, Contemporary Meanings* (New York: Paulist, 2002), 54.

[27] Quoted in Ostling, "Search for the Historical Adam."

[28] Imre Lakatos, "Falsification and the Methodology of Scientific Research Programmes," in *Criticism and the Growth of Knowledge*, ed. Imre Lakatos and Alan Musgrave (Cambridge: Cambridge University Press, 1970), 91-106.

[29] Greg Boyd, "Whether or Not There Was an Historical Adam, Our Faith Is Secure," in Barrett and Caneday, *Four Views on the Historical Adam*, 260.

[30] Ibid., 260.

[31] Ibid., 266.

[32] Martin Luther, "Preface to the Old Testament," in *What Luther Says: An Anthology*, ed. Ewald M. Plass, vol. 1 (St. Louis: Concordia Publishing House, 1959), 71.

[33] Michael Polanyi, *Personal Knowledge: Towards a Post-Critical Philosophy* (Chicago: University of Chicago Press, 1962).

[34] Galileo, "Letter to the Grand Duchess Christina of Tuscany, 1615," Modern History Sourcebook: Galileo Galilei, scitech.au.dk/fileadmin/site_files/science .au.dk/NF/Komm/DenbevaegedeJord/Letter_to_the_Grand_Duchess _Christina_of_Tuscany.1615__Gallilei.pdf.

[35] Harold Nebelsick, "Karl Barth's Understanding of Science," in *Theology Beyond Christendom: Essays on the Centenary of the Birth of Karl Barth, May 10, 1986*, ed. John Thompson (Allison Park, PA: Pickwick, 1986), 201.

[36] William Bruce Cameron, *Informal Sociology: A Casual Introduction to Sociological Thinking* (New York: Random House, 1963), 13.

[37] C. S. Lewis, *Reflections on the Psalms* (New York: HarperCollins, 2017 [1958]), 130.

Case Study: What About Intelligent Design?

[1] Phillip Johnson, *Darwin on Trial* (Downers Grove, IL: InterVarsity Press, 1991).

[2] Perceval Davis and Dean H. Kenyon, *Of Pandas and People: The Central Question of Biological Origins*, 2nd ed. (Mesquite, TX: Haughton, 1993).

[3] For a profound philosophical refutation of ID, see Ric Machuga, *In Defense of the Soul: What It Means to Be Human* (Grand Rapids: Brazos, 2002).

6 Calling Out the Good in Technology

[1] Ray Kurzweil, *The Singularity Is Near* (New York: Penguin, 2005), 136.

[2] Willem B. Drees, "Techno-secularity and Techno-sapiens: Religion in an Age of Technology," *Zygon: Journal of Religion and Science*, February 2013, www.zygon journal.org/technology.html.

[3] Rosalind Picard, *Affective Computing* (Cambridge, MA: MIT Press, 2000), 3.

[4] "Affective Computing, Rosalind W. Picard," MIT Press, accessed August 16, 2017, web.archive.org/web/20080328121832/http://mitpress.mit.edu/catalog /item/default.asp?ttype=2&tid=4060.

[5] Rosalind Picard, MIT personal web page post, April 19, 1995, web.media.mit .edu/~picard/personal/ccc-talk.php.

[6] Rosalind Picard, MIT personal web page post, accessed August 16, 2017, web.media .mit.edu/~picard/personal/faith-test.php.

[7] Jane McGonigal, "Gaming Can Make a Better World," TED, February 2010, www.ted.com/talks/jane_mcgonigal_gaming_can_make_a_better_world.

[8] Jane McGonigal, *SuperBetter: The Power of Living Gamefully* (New York: Penguin, 2016).

[9] I learned this insight in a personal conversation with Margaret Wertheim, author of *The Pearly Gates of Cyberspace: A History of Space from Dante to the Internet* (New York: W. W. Norton, 1999).

[10] Sherry Turkle, *Reclaiming Conversation* (New York: Penguin, 2015), 42.

[11] "Irresistible by Design: It's No Accident You Can't Stop Looking At the Screen," NPR *Fresh Air*, March 13, 2017, www.npr.org/programs/fresh-air/2017 /03/13/519983249/fresh-air-for-march-13-2017.

[12] If this debate interests you, I like how C. E. B. Cranfield sees this as a present Christian experience for Paul in *Romans*, International Critical Commentary, ed. J. A. Emerton and C. E. B. Cranfield (Edinburgh: T&T Clark, 1975).

7 Give Technology a Break

[1] Sherry Turkle, *The Second Self: Computers and the Human Spirit* (Cambridge, MA: MIT Press, 2005); *Alone Together: Why We Expect More from Technology and Less from Each Other* (New York: Basic, 2011); and *Reclaiming Conversation: The Power of Talk in a Digital Age* (New York: Penguin, 2016).

[2] Turkle, *Reclaiming Conversation*, 324.

[3] Ibid., 322.

[4] Ibid., 323.

[5] C. S. Lewis, *The Lion, the Witch, and the Wardrobe* (New York: HarperCollins, 2000 [1950]), 182.

[6] C. G. P. Grey, "Humans Need Not Apply," YouTube, August 13, 2014, www.youtube.com/watch?v=7Pq-S557XQU.

[7] "The human species can, if it wishes, transcend itself. . . . We need a name for this new belief. Perhaps *transhumanism* will serve." Julian Huxley, *Religion Without Revelation* (London: C. A. Watts, 1967), 195, cited in Ted Peters, "Theologians Testing Transhumanism," *Theology and Science* 13, no. 2 (2015): 132.

[8] Nick Bostrom, "Transhumanist Values," *Review of Contemporary Philosophy* 4 (2005): 3, available at www.nickbostrom.com/ethics/values.pdf.

[9] Ted Peters, "Theologians Testing Transhumanism," 147; and "H-: Transhumanism and the Posthuman Future: Will Technological Progress Get Us There?," Metanexus, September 1, 2011, www.metanexus.net/essay/h-transhumanism-and-posthuman-future-will-technological-progress-get-us-there.

[10] Peters, "H-."

[11] Russell Blackford, "H+: Trite Truths About Technology: A Reply to Ted Peters," Metanexus, September 1, 2011, www.metanexus.net/essay/h-trite-truths-about-technology-reply-ted-peters.

[12] Please read Blackford, "H+," for how he exactly phrases it (he's worth reading) and to see what I left out.

[13] Peters, "Theologians Testing Transhumanism," 147.

[14] Christian Transhumanist Association, www.christiantranshumanism.org.

[15] "Pornography Statistics: 250+ Facts, Quotes, and Statistics About Pornography Use," 8, CovenantEyes, 2015, www.covenanteyes.com/resources/download-your-copy-of-the-pornography-statistics-pack.

[16] Ibid.

[17] See Allen B. Downey, "Religious Affiliation, Education and Internet Use," Cornell University Library, March 21, 2014, arxiv.org/abs/1403.5534; and "How the Internet Is Taking Away America's Religion," *MIT Technology Review*, April

4, 2014, www.technologyreview.com/s/526111/how-the-internet-is-taking-away
-americas-religion.

[18]Daniel Dennett, "Why the Future of Religion Is Bleak," *Wall Street Journal*,
April 26, 2015, www.wsj.com/articles/why-the-future-of-religion-is-bleak
-1430104785.

[19]See Robert N. McCauley, *Why Religion Is Natural and Science Is Not* (New York:
Oxford University Press, 2011), 265-68.

[20]Nick Bilton, "Meet Facebook's Mr. Nice," *New York Times,* October 22, 2014,
www.nytimes.com/2014/10/23/fashion/Facebook-Arturo-Bejar-Creating
-Empathy-Among-Cyberbullying.html.

[21]See Bosco Peters, "Virtual Eucharist?," Liturgy, June 28, 2009, liturgy.co.nz
/virtual-eucharist.

[22]Susan Cain, *Quiet: The Power of Introverts in a World That Can't Stop Talking*
(New York: Crown, 2013).

[23]Augustine, *Confessions,* 13.9.10.

[24]Sam Harris, "Can We Build AI Without Losing Control Over It?," TED,
June 2016, www.ted.com/talks/sam_harris_can_we_build_ai_without_losing
_control_over_it.

[25]C. S. Lewis, *The Screwtape Letters* (Grand Rapids: Zondervan, 2001 [1942]), 71.

Two Case Studies: On Global Climate Change and Sexuality

[1]I am leaning on the insights of Lloyd E. Sandelands and Andrew Hoffman,
"Sustainability, Faith, and the Market," *Worldviews* 12 (2008): 129-45, webuser
.bus.umich.edu/lsandel/PDFs/2008%20Worldviews.pdf.

[2]Lynn White Jr., "The Historical Roots of Our Ecologic Crisis," *Journal of the
American Scientific Affiliation* 21 (June 1969): 42-47.

[3]Ibid.

[4]Arnold Toynbee, cited in David R. Kinsley, *Ecology and Religion: Ecological
Spirituality in Cross-Cultural Perspective* (New York: Pearson, 1994), 104.

[5]White, "Historical Roots," 46.

[6]See Greg Cootsona, *Creation and Last Things: At the Intersection of Theology and
Science* (Louisville: Geneva, 2002), chap. 3.

[7]A practical book for ministry leaders, *Creation Care: An Introduction for
Busy Pastors* (2008), is available at www.vineyardresources.com/equip/content
/creation-care.

[8]Cited in Kinsley, *Ecology and Religion*, 122.

[9]Doyle Rice, "Report: Climate Change Is Here and Getting Worse," *USA Today*, May 6, 2014, www.usatoday.com/story/weather/2014/05/06/national-climate-assessment/8736743.

[10]Katharine Hayhoe, "Earthkeeping: A Climate for Change," YouTube, May 13, 2015, www.youtube.com/watch?v=0HbbE74MrUc. See also Katharine Hayhoe and Andrew Farley, *A Climate for Change: Global Warming Facts for Faith-Based Decisions* (New York: FaithWords, 2011).

[11]"Changing Attitudes on Gay Marriage," Pew Research Center Religion & Public Life, June 26, 2017, www.pewforum.org/fact-sheet/changing-attitudes-on-gay-marriage.

[12]Daniel Cox, Juhem Navarro-Riveram, and Robert P. Jones, "A Shifting Landscape: A Decade of Change in American Attitudes About Same-Sex Marriage and LGBT Issues," PRRI, February 26, 2014, www.prri.org/research/2014-lgbt-survey.

[13]American Psychiatric Association, "What Is Gender Dysphoria?," 2016, www.psychiatry.org/patients-families/gender-dysphoria/what-is-gender-dysphoria.

[14]For one source, see the Intersex Society of North America's website, www.isna.org/faq/what_is_intersex.

[15]Alfred C. Kinsey, *Sexual Behavior in the Human Male* (Bloomington: Indiana University Press, 1998); Frank Newport, "Americans Greatly Overestimate Percent Gay, Lesbian in U.S.," Gallup, May 21, 2015, www.gallup.com/poll/183383/americans-greatly-overestimate-percent-gay-lesbian.aspx.

[16]Ted Peters, *Playing God? Genetic Determinism and Human Freedom* (New York: Routledge, 2003), 91-115. First of all, it isn't proven that people become gay by genetic decree. Studying identical twins, who are therefore genetic clones of one another, demonstrates that genetics has a correlation with same-sex attraction, but it's not entirely definitive.

[17]Arthur Peacocke, *Creation and the World of Science*, The Bampton Lectures (Oxford: Clarendon Press, 1979), 112.

[18]See also the CTNS statement on the gay gene discovery in Peters, *Playing God?*, 215-18.

[19]To grasp a sense of diversity within Christian evangelicalism, see, for example, William Loader, Megan K. DeFranza, Wesley Hill, and Stephen R. Holmes, *Two Views on Homosexuality, The Bible, and the Church*, ed. Preston Sprinkle (Grand Rapids: Zondervan, 2016); Wesley Hill, *Washed and Waiting: Reflections on Christian Faithfulness and Homosexuality* (Grand Rapids: Zondervan, 2018); and Gregory Coles, *Single, Gay, Christian* (Downers Grove, IL: InterVarsity Press, 2017).

[20]See, for example, Dr. Joshua Swamidass's sermon "The Science of Sex and Gender," YouTube, June 1, 2017, www.youtube.com/watch?v=eQlHmXsmylU.

[21]"Presbyterian Church (USA) Policy Brief," accessed August 17, 2017, www.pcusa .org/site_media/media/uploads/washington/pdfs/policybrief.pdf.

8 Moving Forward

[1]Andrew Root and Erik Leafblad, "Teaching at the Intersection of Faith and Science," *Youth Worker*, July 1, 2015, www.youthworker.com/youth-ministry -resources-ideas/youth-ministry/11731457. This grant was the youth parallel to my SEYA grant, both funded by the John Templeton Foundation.

[2]David Kinnaman with Aly Hawkins, *You Lost Me: Why Young Christians Are Leaving Church . . . and Rethinking Faith* (Grand Rapids: Baker Books, 2011), 140.

[3]Jonathan Hill, "Do Americans Believe Science and Religion Are in Conflict?," Big Questions Online, April 6, 2015, www.bigquestionsonline.com/2015/04/06 /americans-believe-science-religion-are-conflict.

[4]You can find materials on Scientists in Congregations on the BioLogos website: www.scientistsincongregations.org and https://biologos.org.

[5]Root and Leafblad, "Teaching at the Intersection."

[6]Blaise Pascal, *Pensées*, trans. A. J. Krailsheimer (New York: Penguin, 1966), 83, 85.

[7]Minsu Ha, David L. Haury, and Ross H. Nehm, "Feeling of Certainty: Uncovering a Missing Link Between Knowledge and Acceptance of Evolution," *Journal of Research in Science Teaching*, no. 49 (2011): 95-121.

[8]Jonathan Hill, "National Study of Religion & Human Origins," 2014, available at http://biologos.org/uploads/projects/nsrho-report.pdf.

[9]I'm paraphrasing; see Jonathan Haidt, *The Righteous Mind: Why Good People Are Divided by Politics and Religion* (New York: Vintage, 2013), chap. 3.

[10]Ibid.

[11]C. S. Lewis, *The Collected Letters of C. S. Lewis*, ed. Walter Hooper, vol. 2, *Books, Broadcasts, and the War 1931–1949* (New York: HarperCollins, 2004), 674.

[12]Christian Smith, *Lost in Transition: The Dark Side of Emerging Adulthood* (New York: Oxford University Press, 2011), 241.

[13]Henri Poincaré, *Science and Method*, trans. Francis Maitland (Mineola, NY: Dover, 2003 [1914]), 22.